Lecture Notes in Physics

Volume 993

The series Lecture Notes in Physics (LNP), founded in 1969, reports new developments in physics research and teaching - quickly and informally, but with a high quality and the explicit aim to summarize and communicate current knowledge in an accessible way. Books published in this series are conceived as bridging material between advanced graduate textbooks and the forefront of research and to serve three purposes:

- to be a compact and modern up-to-date source of reference on a well-defined topic;
- to serve as an accessible introduction to the field to postgraduate students and non-specialist researchers from related areas;
- to be a source of advanced teaching material for specialized seminars, courses and schools.

Both monographs and multi-author volumes will be considered for publication. Edited volumes should however consist of a very limited number of contributions only. Proceedings will not be considered for LNP.

Volumes published in LNP are disseminated both in print and in electronic formats, the electronic archive being available at springerlink.com. The series content is indexed, abstracted and referenced by many abstracting and information services, bibliographic networks, subscription agencies, library networks, and consortia.

Proposals should be sent to a member of the Editorial Board, or directly to the responsible editor at Springer:

Dr Lisa Scalone
Springer Nature
Physics
Tiergartenstrasse 17
69121 Heidelberg, Germany
lisa.scalone@springernature.com

More information about this series at http://www.springer.com/series/5304

Gaetano Lambiase • Giorgio Papini

The Interaction of Spin with Gravity in Particle Physics

Low Energy Quantum Gravity

Springer

Gaetano Lambiase
Dipartimento di Fisica E.R. Caianiello
Universita' di Salerno
Fisciano, SA, Italy

Giorgio Papini
Department of Physics
University of Regina
Saskatchewan, SK, Canada

ISSN 0075-8450 ISSN 1616-6361 (electronic)
Lecture Notes in Physics
ISBN 978-3-030-84770-8 ISBN 978-3-030-84771-5 (eBook)
https://doi.org/10.1007/978-3-030-84771-5

This Springer imprint is published by the registered company Springer Nature Switzerland AG
The registered company address is: Gewerbestrasse 11, 6330 Cham, Switzerland

To Laura, Walter, my parents, my sister
and all friends for the constant support
and encouragement
Gaetano Lambiase

To Lucia, my sister Mariella, my sons
and everyone in my family for their patience,
encouragement and affection
Giorgio Papini

Preface

The behaviour of quantum systems in inertial and gravitational fields is of interest in investigations regarding the structure of space–time at the quantum level. Though a definitive answer to questions regarding the fundamental structure of space–time may only come from a successful quantum theory of gravity, the extrapolation of general relativity from planetary lengths, over which it is well established, to Planck length requires a leap of faith in its validity of over forty orders of magnitude and the resolution of difficult quantization problems. The alternative, performing experiments at Planck length, appears, at present, remote. A more realistic approach, limited to verifying Einstein's theory at intermediate lengths, leads necessarily to the study of the interaction of classical inertial and gravitational fields with quantum systems. These include, but are not limited to elementary particles. A number of effects can be predicted in this way with the advantage that Einstein's theory affords a unified treatment of inertia and gravity. A programme of this nature fulfils the double purpose of learning about the reaction of quantum systems to gravity and also of confronting gravitational theory with the precision of quantum measurements. Because of its weakness, gravitation is the fundamental interaction least investigated at atomic and sub-atomic levels.

General relativity incorporates the equivalence principle from the outset and observations, where feasible, do confirm that inertia and gravity interact with quantum systems in ways that are compatible with Einstein's views. This is borne out of measurements on superconducting electrons and on neutrons which offer tangible evidence that the effect of inertia and Newtonian gravity on wave functions is that predicted by wave equations compliant with general relativity down to lengths of 10^{-5} cm and 10^{-8} cm, respectively. General relativity is therefore adhered to in the following, which would hardly require continued reference to inertial fields as distinct entities. Nonetheless, inertial effects must be identified with great accuracy as dictated by their unavoidable presence in Earth-bound and near-space experiments of ever increasing accuracy aimed at testing fundamental theories. In addition, inertia provides a guide in the study of relativity because, in all instances where non-locality is not at stake, the equivalence principle, in some of its forms, ensures the existence of a gravitational effect for each inertial effect.

Quantum particles have mass, charge, and spin. Quantum spin arises from the combination of special relativity with quantum mechanics. Every particle is associated with a particular field and every field corresponds to a class of indistinguishable particles. These belong to two groups, fermions that have half integer spin, and bosons with integer spin. While gravity interacts with mass and charge in a universal way, it interacts with spin in ways that depend essentially on whether a particle is a boson or a fermion. Spin plays an important role in particle physics and is involved in some of the most precise fundamental experiments so far performed. Suffice just to recall here the role of spin in the measurement of the anomalous magnetic moment of the muon. Quantum spins are very versatile tools that can be used in a variety of experimental situations and energy ranges while retaining an essentially non-classical behaviour.

The issue of the interaction of quantum spin with gravity has been vigorously pursued by Mashhoon who has argued that there should be a coupling of rotation, hence gravity, to spin as an extension of the Sagnac effect that describes the coupling of rotation to the angular momentum of a particle. It is with spin that the difference between classical and quantum behaviour of a particle becomes particularly evident.

Fully covariant wave equations incorporate the spin characteristics of an object from the outset. They predict the existence of a class of inertial-gravitational effects that may be tested experimentally in the not too distant future. In these equations inertia and gravity appear as external classical fields, but, by conforming to general relativity, provide valuable information on how Einstein's views carry through in the world of the quantum. The solutions of wave equations to first order in the metric deviation are exact and are characterized by the fact that gravity only appears as a phase in the wave function and can be computed by means of path integrals. It is a Berry phase that can be understood as an effect that gravitation has on Hilbert space. These gravitational Berry phases are derived in Chap. 1 for the most common wave equations and explicitly calculated, in the following chapters, in several instances.

Phase differences are amenable to observation. Interferometers of large dimensions hold great promise in many of these investigations. They can provide accurate measurements of quantum phases, whose role is important in gyroscopy, and in the testing of general relativity. It is anticipated that similar studies will be performed with particle accelerators. The forerunner of this second group of investigations is the work on the measurement of the muon's magnetic moment in which evidence can be found for the coupling of spin to rotation.

The spin-rotation coupling predicted by Mashhoon has been observed directly for photons and neutrons, while applications to neutron interferometry are being developed. This is a most significant development not only because the measurements confirm the existence of the phenomenon, but also because they offer hints of deviation from that universality that people have come to expect of gravity.

Still within the confines of spin physics, the frame dependent spin precession of compound spin systems has become relevant in studies of heavy ions in storage rings. Spintronics has also become a field of practical importance in which inertia, rotational inertia in particular, has a role to play.

For most studies involving quantum particles the gravitational phase is of paramount importance. For other problems, such as the interaction of gravitation at the atomic level, the knowledge of the Hamiltonian is of greater importance. The derivation of the Hamiltonian is usually accomplished by following a sequence of Foldy–Wouthuysen transformations. It has been presented in comprehensive form by Hehl and Ni purely within the framework of special relativity and in the local frame of a fermion. In the present work, the non-relativistic case is tackled by means of a procedure that renders the quantum phase manifest and can, therefore, be applied to a larger class of problems. A high-energy approximation for Dirac particles has been given by Cini and Touschek in 1958. Their work can be extended to include external electromagnetic and gravitational fields and quantum phases to any order. The derivation of the Hamiltonian can then be accomplished in a standard way. Use of some of these results is also made in the External Field Approximation (EFA). The gravitational EFA has developed less rapidly than in the electromagnetic case because of the mathematically involved way in which gravitation is described in Einstein's theory. It is, however, a useful tool in practical applications. It is developed and applied in Chaps. 1 and 2.

Gravitational-inertial fields in the laboratory are weak and range from weak to intermediate strength for most astrophysical sources. These are the fields considered here. They are adequately described by the Weak Field Approximation (WFA) and can be applied, with caution, even in the neighbourhood of a strong source if $GM/c^2 r \leq 10^{-1}$.

Of note among the applications are those to non-relativistic superconductors by DeWitt, whose results were obtained when it was unknown whether general relativity could be applied in a quantum mechanical context. Results that applied to non-inertial fields could find experimental confirmation, in an appropriate limit, in the London moment of rotating superconductors.

Applications to interferometry as a way to test the action of gravity on the wave function of quantum systems are also given in Chap. 3.

Chapter 4 is devoted to the behaviour of neutrinos and their helicity and flavour oscillations in gravitational fields and to some consequences like spin coupling to primordial gravitational waves and a pulsar's increase in its rotation rate (pulsar kick). Chapter 5 also deals with neutrinos, in particular with the effect of general relativity on neutrino spin flavour and spin-flip oscillations above the core of type II supernovae and off rotating black holes.

Some particular radiative processes are considered in Chap. 6 together with the spin currents generated by non-inertial fields in which, unlike non-relativistic particles, spin and orbital angular momenta are not separately conserved. Still in Chap. 6, relativistic vortices are shown to arise as solutions of the covariant Dirac equation.

Chapter 7 is concerned with axion electrodynamics and the search for axions as leading candidates for dark matter, with their influence on angular momentum and spin of particles and with the formation of a chirally active space–time medium. Current ideas on geometrical developments like torsion are also introduced.

Chapter 8 deals with perspectives on several arguments treated in the book. The focus is mainly on the Gravitational Memory Effect and on the Quasi Normal Modes, which could be detected with a future generation of gravitational wave detectors. In addition, some important developments in holography, double-copy, and Kawai–Lewellen–Tye relations, together with a recent proposal concerning spin entanglement as a witness for quantum gravity are briefly discussed.

Spin-inertia and spin-gravity interactions are the subject of numerous theoretical and experimental efforts. An extensive, but hardly complete bibliography, is attached.

Salerno, Italy Gaetano Lambiase
Toronto, Canada Giorgio Papini
June 2021

Contents

1 Quantum Systems in Gravitational Fields. Berry Phases . . . 1
1.1 Introduction . . . 1
1.2 The Effect of Space–Time Curvature on Hilbert Space. Berry Phase . . . 2
1.3 Wave Equations . . . 6
1.3.1 The Schroedinger Equation. Superfluids . . . 7
1.3.2 The Klein–Gordon Equation . . . 8
1.3.3 Spin-1 Equations . . . 10
1.3.4 The Dirac Equation in Curved Space–Time . . . 12
1.3.5 The Lense–Thirring Metric . . . 16
1.3.6 Spin-2 Particles in Gravitational Fields . . . 20
Problems . . . 26
References . . . 26

2 The Mashhoon Effect. Spin-Gravity Interactions . . . 29
2.1 Introduction . . . 29
2.2 Spin-Rotation Coupling . . . 30
2.2.1 Spin-Rotation Coupling in Muon g-2 Experiments . . . 31
2.2.2 Spin-Rotation Coupling and Limits on *P* and *T* Invariance . . . 33
2.3 Spin-Rotation Coupling in Compound Spin Objects . . . 37
2.4 Helicity Precession of Fermions in Gravitational Fields . . . 42
2.5 Chirality Precession of Fermions in Gravitational Fields . . . 46
Problems . . . 48
References . . . 48

3 Interferometers in Gravitational Fields . . . 51
3.1 Introduction . . . 51
3.1.1 Superconductors . . . 52
3.1.2 Gravitational Waves and Superfluids . . . 53

3.2 Interferometers in Various Metrics 55
3.2.1 Interferometer in the Field of Earth 55
3.2.2 Rotation 57
3.2.3 The Lense–Thirring Effect for Quantum Systems 58
3.3 Wave Optics 60
Problems 65
References 66

4 Neutrinos in Gravitational Fields 69
4.1 Introduction 69
4.1.1 Neutrino Helicity Oscillations 69
4.1.2 Helicity Oscillations in a Medium 72
4.1.3 Neutrino Flavour Oscillations: The Effect of Neutrino's Travel Time 73
4.2 Neutrino Optics 74
4.3 Helicity Transitions Induced by Gravitational Fields 77
4.4 Neutrino Flavour Oscillations 81
4.5 Neutrino Lensing 90
4.6 Spin-Gravity Coupling Of Neutrinos with Primordial Gravitational Waves 92
4.7 Pulsar Kick 93
Problems 94
References 95

5 Neutrinos Physics: Further Topics 97
5.1 Introduction—The Standard Model 97
5.2 General Aspects of Neutrino Spin-Flavour and Spin-Flip 101
5.2.1 The Equation of Evolution in Curved Space–Time 102
5.2.2 Relativistic Effects Near a Static Star 104
5.2.3 Relativistic Effects Near a Rotating Star 105
5.3 Spin Oscillations of Neutrinos Scattered Off a Rotating Black Hole 106
5.4 Lensing and Oscillations Probability 109
5.4.1 Gravitational Lensing of Neutrinos in Schwarzschild Space–Time 109
References 111

6 Radiative Processes, Spin Currents, Vortices 113
6.1 Radiative Processes 113
6.2 Spin Currents in Gravitational Fields 121
6.3 Vortices 125
6.3.1 Spin-$\frac{1}{2}$ Fermions 129

6.4 Zitterbewegung and Gravitational Berry Phase 130
Problems 134
References 134

7 Other Developments 137
7.1 Scalar–Pseudoscalar Coupling and the Search for Axions 137
7.2 Axion Electrodynamics 139
7.3 The Extended Bargmann–Michel–Telegdi Model 139
7.4 Space–Times with Torsion 141
7.4.1 Spin-Flip Transitions in Space–Times with Torsion 141
7.4.2 More About Torsion 142
7.5 Axions and Berry Phase 143
Problems 147
References 148

8 Perspectives 151
8.1 The Gravitational-Wave Memory Effect 151
8.2 Quantum Wave Equations and Quasi-normal Modes 159
8.3 Conclusions 163
References 165

9 Conclusions 169
References 171

A Natural Units and Conventions 173

B Dirac Matrices 175

C Neutrino Oscillations in Flat and Curved Space-Time 177

D Dirac Hamiltonians in the Low-Energy Approximation 179

E Fermion Helicity Flip in Weak Gravitational Fields 185

Index 189

Acronyms[1]

BMT Bargmann Michel Telegdi
EFA External Field Approximation
KG Klein Gordon
MSW Mechanism
QED Quantum Electrodynamics
QNMs Quasi Normal Modes
SM Standard Model
WFA Weak Field Approximation
ZB Zitterbewegung

[1]Lists of abbreviations, symbols and the like are easily formatted with the help of the Springer-enhanced description environment.

Quantum Systems in Gravitational Fields. Berry Phases

1

Abstract

This chapter deals with Berry phase and its extension to relativistic quantum systems represented by known wave equations. Solutions that are exact to first order in the metric deviation $\gamma_{\mu\nu}$ are given for Klein–Gordon, Maxwell–Proca, Dirac and spin-2 equations. The corresponding Berry phases are expressed in terms of $\gamma_{\mu\nu}$.

1.1 Introduction

Covariant wave equations are particularly useful in dealing with quantum systems because they contain information about the systems they describe and can be promptly derived from their flat space–time counterparts, thanks to the equivalence principle, by simply replacing the Minkowski metric $\eta_{\mu\nu}$ with $g_{\mu\nu}$. A notable exception to this rule is represented by the Dirac equation because, as explained below, there are no representations of the group $GL(4, R)$ that behave like spinors under $SO(3, 1)$. A covariant equation for spin-1/2 particles can nonetheless be derived.

Though non-covariant, the Schroedinger equation is also considered because of its applicability to non-relativistic quantum systems of relevant experimental interest. The other wave equations considered are the Klein–Gordon equation for spin-0 particles, the Maxwell and Proca equations for massless and massive spin-1 particles respectively and the massless and massive spin-2 wave equations (in the massive case the equation goes under the name of Fierz–Pauli). The solutions given are all exact to first order in the metric deviation $\gamma_{\mu\nu} = g_{\mu\nu} - \eta_{\mu\nu}$, can be extended to any order in it, but there does not seem to be a general summation procedure. The solution of covariant wave equations yields in general meaningful insights into aspects of the interaction of quantum systems with gravity whenever the gravitational field need not be quantized.

G. Lambiase and G. Papini, *The Interaction of Spin with Gravity in Particle Physics*, Lecture Notes in Physics 993, https://doi.org/10.1007/978-3-030-84771-5_1

Fully covariant wave equations predict the existence of a class of inertial-gravitational effects that can be tested experimentally. In these equations inertia and gravity appear as external classical fields, but, by conforming to general relativity, provide information on how Einstein's theory behaves in the world of the quantum. Experiments already confirm that inertia and Newtonian gravity affect quantum particles in ways that are fully consistent with general relativity down to distances of $\sim 10^{-5}$ cm for superconducting electrons [1] and of $\sim 10^{-8}$ cm for neutrons [2–5]. Other aspects of the interaction of gravity with quantum systems are just beginning to be investigated.

Gravitational fields affect particle wave functions in a variety of ways. They induce quantum phases that afford a unified treatment of interferometry and gyroscopy. They interact with particle spins giving rise to a number of significant effects. They finally shift energy levels in particle spectra [6,7]. While it still is difficult to predict when direct measurements will become possible in the latter case, rapid experimental advances in particle interferometry [8–10] require that quantum phases be derived with precision. This is done below for Schroedinger, Klein–Gordon, Maxwell and Dirac equations. The development of large, sensitive interferometers appears essential in many of these investigations. They can play a role in the testing of general relativity and of fundamental theories [11,12] in Earth-bound and near space experiments [13].

Spin-inertia and spin-gravity interactions have been the subject of several theoretical [14–24] and experimental efforts [25–31]. It is shown below that spin-rotation coupling is particularly important in precise tests of fundamental theories and in certain types of neutrino oscillations. Particle accelerators may be also called to play a role in these investigations [32,33].

1.2 The Effect of Space–Time Curvature on Hilbert Space. Berry Phase

The line of reasoning followed below can be simply illustrated for scalar particles but can be generalized to include particles of any spin.

General relativity describes gravitational effects in space–time and explains gravitational interactions as space–time geometry. Quantum mechanics, on the contrary, employs Hilbert space as the representation space for quantum systems. The geometrical structure of Hilbert space has however remained largely unnoticed until the discovery of Berry phase [34]. The parallel transport of the eigenvectors of a quantum system whose Hamiltonian $H(\lambda)$ depends on a set of slowly varying parameters λ_a leads in fact to a geometrical phase $\gamma(C)$ for every transport path C in the space of the λ's. The evolution of the system is also accompanied by the appearance of Abelian gauge potentials [35] or non-Abelian ones [36] when degeneracy is present. Thus, the gauge potentials, which represent the geometrical structure of parameter space, induce a corresponding structure in Hilbert space which results in Berry's phase. Since gravitational fields endow space–time with a geometrical structure, the latter should also affect Hilbert space *when parameter space and space–time coincide*. This is shown below for the simpler, but physically important case of weak gravitational fields.

The identification of space–time with parameter space, dictated by relativity, requires a fully covariant description of the evolution of the quantum system itself. Thus, the theory of Berry's phase, which is based on the non-relativistic Schrödinger equation, must be made covariant. This is accomplished by using the proper time formalism of Fock, Nambu, Feynman, and Schwinger [37]. With this generalization, the gravitational gauge field can be derived from a covariant geometrical phase and defined in space–time. The results of Ref. [38–40] can also be extended to nonlinear systems [40] by following the work of Garrison and Chiao [41] and Anandan [42].

As shown by Berry [34], the parallel transport of the eigenvectors of a quantum system whose Hamiltonian $H(\lambda)$ depends on a set of slowly varying parameters λ_a leads to a geometrical phase $\gamma(C)$ for every transport path C in the space of the λ's.

Linear systems are only considered below. For scalar particles in the presence of some arbitrary interactions which can be described in terms of a potential $U(x)$, the Klein–Gordon equation is

$$(\Box + m^2)\phi(x) = U(x)\phi(x). \tag{1.1}$$

With the help of the appropriate Green's function $G(x, x')$, Eq. (1.1) together with Feynman's boundary conditions [43] can be transformed into the integral equation

$$\phi(x) = e^{-ik\cdot x} - \int d^4x' G(x, x')U(x')\phi(x'). \tag{1.2}$$

Equation (1.1), however, can be also re-cast into the form of a Schrödinger equation

$$i\frac{\partial \Phi}{\partial \tau} = H\Phi \tag{1.3}$$

by introducing the proper time τ, a relativistic invariant parameter which describes the evolution generated by the relativistic Hamiltonian

$$H \equiv -(\Box - U(x)). \tag{1.4}$$

The relation between the variables x^μ and τ is fixed by assuming $\Phi(x, \tau) = e^{-im^2\tau}\phi(x)$, where $\phi(x)$ satisfies the equation

$$H\phi(x) = -m^2\phi(x), \tag{1.5}$$

which coincides with (1.1). Since, in general, $U(x)$ does not depend explicitly on τ, the type of solution chosen for (1.3) does indeed exist [38]. Equation (1.3) can also be applied to vector and spinor fields [38].

The covariant generalization of Berry's phase [38] can be obtained by applying Berry's argument [35] to Eq. (1.3) with a parameter-dependent Hamiltonian, i.e.

$$i\frac{\partial \Phi}{\partial \tau} = H(\lambda_a(\tau))\Phi, \qquad a = 1, 2, \ldots k, \tag{1.6}$$

and by replacing the condition that the evolution be adiabatically slow with the assumption that

$$\sum_{m' \neq m''} \left| \frac{\langle m'|\frac{\partial H}{\partial \tau}|m''\rangle}{m'^2 - m''^2} \right|^2 \ll 1, \tag{1.7}$$

where

$$\frac{\partial H}{\partial \tau} \equiv \frac{\partial H}{\partial \lambda_a} \frac{d\lambda_a}{d\tau}, \tag{1.8}$$

and $|m'\rangle$ denotes an instantaneous eigenstate of H with eigenvalue m'^2. After a cycle T, one has

$$\Phi(x, \lambda_a(T)) = \exp(i\gamma(C)) \exp(-im^2 T) \Phi(x, \lambda_a(0)) \tag{1.9}$$

with a corresponding phase (see Fig. 1.1)

$$\gamma(C) = i \oint_c \int \phi^* \frac{\partial}{\partial \lambda_a} \phi d\lambda_a d^4 x. \tag{1.10}$$

Finally, one obtains

$$\phi(x, \lambda_a(T)) = \exp(i\gamma(C))\phi(x, \lambda_a(0)), \tag{1.11}$$

where $A^a \equiv \int \phi^* \frac{\partial}{\partial \lambda_a} \phi d^4 x$, defined in the k-dimensional parameter space, is the gauge field of Ref. [36]. The main point of the covariant generalization given is represented by the fact that it is now possible to identify parameter space with ordinary space–time by setting $k = 4$. The parameters are then the collective coordinates of the system considered and A^a becomes a gauge field in the usual sense. Specifically, these collective coordinates could be the coordinates of some particular point of the system, e.g. the centre of mass, and are different from the x^μ which may be regarded as relative coordinates with respect to the collective ones.

In the weak-field approximation one can rewrite the covariant Klein–Gordon equation

$$(\nabla_\mu \nabla^\mu + m^2)\phi = 0 \tag{1.12}$$

in the form of (1.1). To order $O(\gamma_{\mu\nu})$, one obtains

$$(\Box + m^2)\phi(x) = U(x)\phi(x), \tag{1.13}$$

with

$$U(x) \equiv \gamma_{\mu\nu} \partial^\mu \partial^\nu - \left(\frac{1}{2} \gamma_\mu{}^\mu{}_{,\nu} - \gamma_\nu{}^\mu{}_{,\mu} \right) \partial^\nu. \tag{1.14}$$

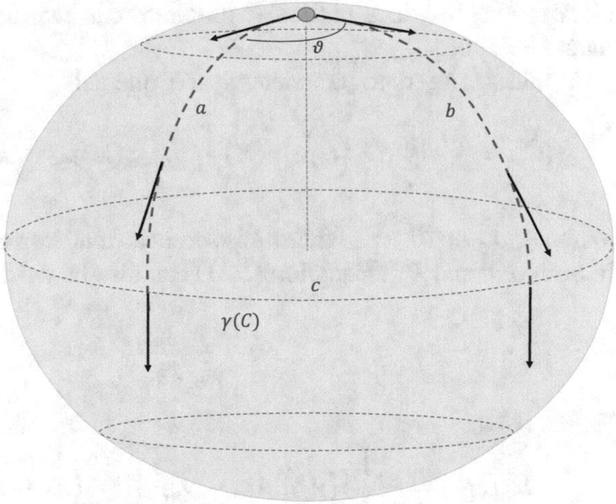

Fig. 1.1 Geometric phase acquired by transporting a vector (state) by parallel displacement along a closed loop $\gamma(C)$ formed by the curves a, b and c. The vector is rotated by the angle ϑ

By using Green's function $G(x, x')$ for (1.13), one obtains

$$\phi(x) = \phi_0(x) + \int d^4x' G(x, x')U(x')\phi(x'), \tag{1.15}$$

where ϕ_0 satisfies the field-free equation

$$(\Box + m^2)\phi_0 = 0. \tag{1.16}$$

In the first-order approximation, $\phi(x')$ in (1.15) can be replaced by $\phi_0(x')$. By substituting the expression of $U(x)$ into (1.15) and integrating by parts, one arrives at the first-order solution

$$\phi(x) = e^{-i\chi}\phi_0(x), \tag{1.17}$$

where

$$\chi \equiv -\frac{1}{4}\int_X^x dz^\lambda(\gamma_{\alpha\lambda,\beta}(z) - \gamma_{\beta\lambda,\alpha}(z))\hat{L}^{\alpha\beta}(z) + \frac{1}{2}\int_X^x dz^\lambda \gamma_{\alpha\lambda}(z)\hat{P}^\alpha. \tag{1.18}$$

The integration is taken along a path from the arbitrary point X to the field point x and the operators $\hat{L}^{\alpha\beta}(z)$ and $\hat{P}^\alpha$ are defined by

$$\begin{aligned}[\hat{L}^{\alpha\beta}(z), \phi_0(x)] &= \hat{L}^{\alpha\beta}(z)\phi_0(x) \equiv \\ &\equiv i\left[(x^\alpha - z^\alpha)\partial^\beta\phi_0(x) - (x^\beta - z^\beta)\partial^\alpha\phi_0(x)\right]\end{aligned} \tag{1.19}$$

$$[\hat{P}^\alpha, \phi_0(x)] = \hat{P}^\alpha\phi_0(x) \equiv i\partial^\alpha\phi_0(x). \tag{1.20}$$

It can be verified that this solution is also valid for the Landau–Ginzburg equation. It should be pointed out that the integration path in (1.18) is arbitrary. However when

(1.18) is applied to a particular problem, e.g. interferometry, the integration path must be defined.

When X^μ refers to parameter space, one gets

$$\gamma(C) = -\frac{1}{4}\oint_C dX^\lambda(\gamma_{\alpha\lambda,\beta}(X) - \gamma_{\beta\lambda,\alpha}(X))L^{\alpha\beta}(X) + \frac{1}{2}\oint_C dX^\lambda \gamma_{\alpha\lambda}(X)P^\alpha . \tag{1.21}$$

When $\phi_0 = Ae^{-ik_\mu X^\mu}$, where A is some normalizing constant, ∂^α is replaced by ik^α in both $J^{\alpha\beta}$ and P^α. Equation (1.21) can also be written as

$$\gamma(C) = \oint dz^\lambda K_\lambda(z, x) , \tag{1.22}$$

where

$$K_\lambda(z, x) = -\frac{1}{2}\left[\left(\gamma_{\alpha\lambda,\beta}(z) - \gamma_{\beta\lambda,\alpha}(z)\right)\left(x^\alpha - z^\alpha\right) - \gamma_{\beta\lambda}(z)\right]k^\beta \tag{1.23}$$

is the gauge field generated by Berry's phase. Notice the presence of the additional parameter k^β that refers to the free particle. The new gauge field represents in essence a quasiparticle.

By using Stokes' theorem, (1.21) becomes

$$\gamma(C) = \frac{1}{4}\int_{\Sigma_c} R_{\mu\nu\alpha\beta}L^{\alpha\beta}d\sigma^{\mu\nu}, \tag{1.24}$$

where $R_{\mu\nu\alpha\beta}$ is the linearized Riemann curvature tensor

$$R_{\mu\nu\alpha\beta} = \frac{1}{2}\left(\gamma_{\mu\beta,\nu\alpha} + \gamma_{\nu\alpha,\mu\beta} - \gamma_{\mu\alpha,\nu\beta} - \gamma_{\nu\beta,\mu\alpha}\right) . \tag{1.25}$$

The introduction of electromagnetic fields is trivial and results in the addition of the term $\int F_{\mu\nu}d\sigma^{\mu\nu}$ to $\gamma(C)$. The integration path C is now determined by the evolution curve of the system in space–time. The solutions of the covariant wave equations given in the following can therefore be derived from the geometrical structure of Hilbert space. They are expressible in terms of genuine quantum phases generated by the gauge fields (1.23). This proves the consistency of the results obtained.

Feng and Lee have calculated Berry's phase for the Kerr–Schild metric and found that the effect is analogous to the optical rotation of a linearly polarized light propagating in a helical optical fibre [44].

Bai and Wang have studied the Berry phase of a photon in the gravitational field of a high-power laser [45].

1.3 Wave Equations

The quantum phases induced by gravity are derived in this section for Schroedinger, Klein–Gordon, Maxwell–Proca, Dirac and Fierz–Pauli equations. Some applications are given in Chap. 2.

1.3.1 The Schroedinger Equation. Superfluids

Though non-covariant, the Schroedinger equation can be applied to the study of some interesting systems, like quantum fluids, that offer, in principle, the possibility of experimental applications. In view of this, it is convenient to start from the action principle

$$S = -m\int ds = -m\int \sqrt{g_{\mu\nu}\dot{x}^{\mu}\dot{x}^{\nu}}dx^0, \tag{1.26}$$

where $\dot{x}^{\mu} = dx^{\mu}/dx^0$, which yields the Lagrangian

$$L = -m(g_{ij}\dot{x}^i\dot{x}^j + 2g_{i0}\dot{x}^i + g_{00})^{1/2}, \tag{1.27}$$

where $i, j = 1, 2, 3$. The momenta p_i are obtained from L by using their definition

$$p_i = \partial L/\partial \dot{x}^i = -m(g_{ij}\dot{x}^j + g_{i0})(g_{lk}\dot{x}^l\dot{x}^k + 2g_{k0}\dot{x}^k + g_{00})^{-1/2}. \tag{1.28}$$

and, after substitution into $H = p_i\dot{x}^i - L$, yield the Hamiltonian

$$H = m(g_{i0}\dot{x}^i + g_{i0})(g_{lk}\dot{x}^l\dot{x}^k + 2g_{k0}\dot{x}^k + g_{00})^{-1/2}. \tag{1.29}$$

Using the WFA, $g_{\mu\nu} \simeq \eta_{\mu\nu} + \gamma_{\mu\nu}$, $g^{\mu\nu} \simeq \eta^{\mu\nu} - \gamma^{\mu\nu}$, $g^{ij}g_{jk} \simeq \delta^i_k$, one obtains from (1.28) the result

$$g_{lk}\dot{x}^l\dot{x}^k + 2g_{k0}\dot{x}^k + g_{00} = \frac{g_{00} - g^{il}g_{i0}g_{l0}}{1 - \frac{1}{m^2}g^{lk}p_l p_k}, \tag{1.30}$$

which, substituted into (1.29), gives

$$H \simeq \sqrt{p^2 + m^2}\left(1 + \frac{1}{2}\gamma_{00}\right) + \frac{1}{2\sqrt{p^2 + m^2}}\gamma^{ij}p_i p_j - p^l\gamma_{l0}. \tag{1.31}$$

In the presence of electromagnetic fields and in the low velocity limit, the Hamiltonian (1.31) leads to the Schroedinger equation [46,47]

$$i\frac{\partial\psi(x)}{\partial t} = \left[\frac{1}{2m}(p_i - eA_i + m\gamma_{0i})^2 - eA_0 + \frac{1}{2m}\gamma_{00}\right]\psi(x). \tag{1.32}$$

The WFA does not fix the reference frame entirely. The transformations $x_\mu \to x_\mu + \xi_\mu$, with ξ_μ small, are still allowed and lead to the "gauge" transformations

$$\gamma_{\mu\nu} \to \gamma_{\mu\nu} - \xi_{\mu,\nu} - \xi_{\nu,\mu}\,.$$

In the *stationary* case, the transformations $\gamma_{00} \to \gamma_{00}, \gamma_{0i} \to \gamma_{0i} - \xi_{0,i}$ leave Eq. (1.32) gauge invariant. Returning to normal units, the solution of the Schroedinger equation is, in this case,

$$\psi(x) = \exp\left\{i\frac{mc}{\hbar}\int^{x}\gamma_{0i}dx^{i} - i\frac{e}{c\hbar}\int^{x}A_{i}dx^{i}\right\}\psi_{0}(x), \tag{1.33}$$

where ψ_0 is the solution of the field-free Schroedinger equation.

If the electron-lattice interaction is added to Eq. (1.32), then the resulting equation can be applied to the study of superconductors in weak stationary gravitational fields [46,47]. Beside behaving as charged, non-viscous fluids, superconductors also exhibit quantization on a macroscopic scale and appear ideally suited to measure small physical effects. The literature on the subject covers various aspects of the behaviour and uses of superconductors and superfluids in gravitational fields [48–51].

Bound states of ultra cold neutrons in Earth's gravitational field

By allowing ultra cold neutrons to fall towards a horizontal mirror, in the gravitational field of Earth, the falling neutrons do not move continuously, but jump from a height to another, as predicted by quantum mechanics [52–54]. The horizontal mirror and Earth's gravitation provide the potential well in which the neutrons are confined. The standing waves created by the reflection of neutrons by the mirror have maxima and minima along the vertical direction whose positions depend on the quantum number of the bound state according to the Bohr–Sommerfeld formula

$$E_n = \sqrt[3]{\frac{9m\pi\hbar g}{8}\left(n - \frac{1}{4}\right)}. \tag{1.34}$$

The dropped neutrons go through quantum states, an additional proof that quantum mechanics also applies when gravitational fields are present. The lowest quantum state has an energy 1.41×10^{-12} eV, approximately equal to the energy needed to lift a neutron by 10 μm in the gravitational field of Earth [53], which makes the experimental method used ideally suited to test Newton's law at the micrometre scale and to place constraints on small effects due to dark matter and dark energy [54].

1.3.2 The Klein–Gordon Equation

A well-known form of the fully covariant Klein–Gordon equation is

$$\left(g^{\mu\nu}\nabla_{\mu}\nabla_{\nu} + m^{2}\right)\Phi(x) = 0, \tag{1.35}$$

where ∇_μ represents covariant differentiation. To first order in WFA, Eq. (1.35) becomes

$$\left[(\eta^{\mu\nu} - \gamma^{\mu\nu})\partial_\mu\partial_\nu - \left(\gamma^{\alpha\mu} - \frac{1}{2}\gamma^\sigma_\sigma\eta^{\alpha\mu}\right)_{,\mu}\partial_\alpha + m^2\right]\Phi(x) = 0, \tag{1.36}$$

and has the exact solution [55,56]

$$\Phi(x) = e^{-i\Phi_G}\phi_0(x) = (1 - i\Phi_G)\phi_0(x), \tag{1.37}$$

where $\phi_0(x)$ is the solution of the field-free equation in Minkowski space, and

$$i\Phi_G\,\phi_0 = \left\{\frac{1}{4}\int_P^x dz^\lambda \left(\gamma_{\alpha\lambda,\beta}(z) - \gamma_{\beta\lambda,\alpha}(z)\right)\left[(x^\alpha - z^\alpha)\partial^\beta - (x^\beta - z^\beta)\partial^\alpha\right] - \right.$$
$$\left. - \frac{1}{2}\int_P^x dz^\lambda\gamma_{\alpha\lambda}(z)\partial^\alpha\right\}\phi_0. \tag{1.38}$$

It follows from Sect. 1.2 that Eq. (1.38) represents Berry's phase [34]. It is easy to prove by direct substitution and using the Lanczos–DeDonder gauge condition

$$\left(\gamma^{\beta\mu} - \frac{1}{2}\gamma^\sigma_\sigma\eta^{\beta\mu}\right)_{,\mu} = 0, \tag{1.39}$$

that (1.37) is a solution of (1.36). For a closed path in space–time one finds

$$i\Delta\Phi_g\phi_0 = \frac{1}{4}\oint R_{\mu\nu\alpha\beta}L^{\alpha\beta}d\tau^{\mu\nu}\phi_0, \tag{1.40}$$

where $L^{\alpha\beta}$ is the angular momentum of the particle of mass m and $R_{\mu\nu\alpha\beta}$ is the linearized Riemann tensor (1.25). The result found is therefore manifestly gauge invariant.

Unlike the case of the Schroedinger equation discussed above, the gravitational fields considered in this section need not be stationary.

Equation (1.37) may also be applied to describe relativistic charged and neutral superfluids and Bose–Einstein condensates.

The solution (1.37) and (1.38) can be iterated to any order n using the relation

$$\phi(x) = \sum_n \phi_{(n)}(x) = \sum_n e^{-i\hat{\Phi}_G}\phi_{(n-1)}, \tag{1.41}$$

where $\hat{\Phi}_G\phi_0 = \Phi_G\phi_0$.

Additional applications of (1.37) are discussed below [55–57].

Corichi and Pierri have studied the Berry phase acquired by a scalar particle transported along a closed path linking a rotating cosmic string [58], while the phase of a scalar particle in the presence of a magnetic cosmic string has been calculated by Mostafazadeh [59].

Berry's phase in the presence of axions and torsion will be derived in Chap. 7.

Gravitational redshift

The simplest possible application of (1.37) is to the calculation of the gravitational redshift. Consider two light sources of the same frequency placed at distances r_A and r_B from the origin at the initial time x_1^0. They are compared at r_A at the later time x_2^0. Neglecting spin effects, the phase difference can be simply obtained from (1.38) using the closed space–time path in the (r, x^0)-plane with vertices at (r_A, x_1^0), (r_B, x_1^0), (r_B, x_2^0), (r_A, x_2^0). The gravitational field is represented by $\gamma_{00}(r) = 2\varphi(r)$, where $\varphi(r)$ is the Newtonian potential. One finds

$$
\begin{aligned}
\Delta\chi &= \frac{1}{2}\int_{x_1^0}^{x_2^0} dz^0[\gamma_{\alpha 0,\beta}(r_B) - \gamma_{\beta 0,\alpha}(r_B)](x^\alpha - z^\alpha)k^\beta + \\
&+\frac{1}{2}\int_{x_2^0}^{x_1^0} dz^0\gamma_{\alpha 0,\beta}(r_A)(x^\alpha - z^\alpha)k^\beta - \frac{1}{2}\int_{x_1^0}^{x_2^0} dz^0\gamma_{\alpha 0}(r_B)k^\alpha - \\
&-\frac{1}{2}\int_{x_2^0}^{x_1^0} dz^0\gamma_{\alpha 0}(r_A)k^\alpha = \\
&= -\frac{k^0}{2}(x_2^0 - x_1^0)[\gamma_{00}(r_B) - \gamma_{00}(r_A)] - \frac{k^0}{4}(x_2^0 - x_1^0)^2[\gamma_{00,1}(r_B) + \gamma_{00,1}(r_A)] .
\end{aligned}
\tag{1.42}
$$

The first term gives the usual redshift formula

$$
\left(\frac{\Delta\nu}{\nu}\right)_1 = -\frac{1}{c^2}[\varphi(r_B) - \varphi(r_A)] .
$$

The second term yields the additional correction

$$
\left(\frac{\Delta\nu}{\nu}\right)_2 = -\frac{x_2^0 - x_1^0}{4}[\gamma_{00,1}(r_B) + \gamma_{00,1}(r_A)] .
$$

In an experiment of the type carried out by Pound and Rebka the ratio of the two terms is $(\frac{\Delta\nu}{\nu})_2/(\frac{\Delta\nu}{\nu})_1 \simeq \frac{2l}{R_\oplus}$, where $l = r_B - r_A$. The second term $(\frac{\Delta\nu}{\nu})_2$ becomes larger at sufficiently high values of l.

1.3.3 Spin-1 Equations

Consider Maxwell equations

$$
\nabla_\nu\nabla^\nu A_\mu - R_{\mu\sigma}A^\sigma = 0, \tag{1.43}
$$

where the electromagnetic field A_μ satisfies the condition $\nabla_\mu A^\mu = 0$. If the second term in Eq. (1.43) is negligible, then Maxwell equations in the WFA are

$$
\nabla_\nu\nabla^\nu A_\mu \simeq (\eta^{\sigma\alpha} - \gamma^{\sigma\alpha})A_{\mu,\alpha\sigma} - (\gamma_{\sigma\mu,\nu} + \gamma_{\sigma\nu,\mu} - \gamma_{\mu\nu,\sigma})A^{\sigma,\nu} = 0, \tag{1.44}
$$

where use has been made of the Lanczos–DeDonder gauge condition (1.39). Equation (1.44) has the solution [56]

$$A_\mu(x) = a_\mu(x) - \frac{1}{4}\int_P^x dz^\lambda \left(\gamma_{\alpha\lambda,\beta}(z) - \gamma_{\beta\lambda,\alpha}(z)\right)[(x^\alpha - z^\alpha)\partial^\beta a_\mu(x) -$$
$$-(x^\beta - z^\beta)\partial^\alpha a_\mu(x)] + \frac{1}{2}\int_P^x dz^\lambda \gamma_{\alpha\lambda}(z)\partial^\alpha a_\mu(x) +$$
$$+\frac{1}{2}\int_P^x dz^\lambda(\gamma_{\beta\mu,\lambda}(z) + \gamma_{\beta\lambda,\mu}(z) - \gamma_{\mu\lambda,\beta}(z))a^\beta(x), \quad (1.45)$$

where $\partial_\nu\partial^\nu a_\mu = 0$ and $\partial^\nu a_\nu = 0$. Equation (1.45) can be written in the form $A_\mu = exp(-i\xi)a_\mu$, where

$$\xi = -\frac{1}{4}\int_P^x dz^\lambda(\gamma_{\alpha\lambda,\beta}(z) - \gamma_{\beta\lambda,\alpha}(z))J^{\alpha\beta} +$$
$$+\frac{1}{2}\int_P^x dz^\lambda \gamma_{\alpha\lambda}(z)k^\alpha - \frac{1}{2}\int_P^x dz^\lambda \gamma_{\alpha\beta,\lambda}(z)T^{\alpha\beta}, \quad (1.46)$$

$$(S^{\alpha\beta})_{\mu\nu} = -i(g^{\mu\alpha}g^{\nu\beta} - g^{\mu\beta}g^{\nu\alpha})$$

is the spin-1 operator, and

$$(T^{\alpha\beta})^{\mu\nu} \equiv -i\frac{1}{2}(g^{\mu\alpha}g^{\nu\beta} + g^{\mu\beta}g^{\nu\alpha}),$$

with k^α the momentum of the free photon and $k_\alpha k^\alpha = 0$ and a_μ, on which ξ operates, is considered as a one-column, four-rows matrix. All spin effects are therefore contained in the $S^{\alpha\beta}$ term [60]. For a closed path one finds (1.40), where the total angular momentum $J_{\alpha\beta} = L_{\alpha\beta} + S_{\alpha\beta}$ now substitutes $L_{\alpha\beta}$. Higher order approximations can be derived using the iteration procedure (1.41).

Spin-gravity coupling for spin-1 particles

The spin-rotation coupling derived by Mashhoon by extending the hypothesis of locality can be now derived rigorously from the solutions found.

Applying the time integral part of ξ

$$\xi = -\frac{1}{4}\int_p^x dz^0 \left(\gamma_{\alpha0,\beta}(z) - \gamma_{\beta0,\alpha}(z)\right) S^{\alpha\beta} - \frac{1}{2}\int_p^x dz^0 \gamma_{\alpha\beta,0}(z)T^{\alpha\beta}, \quad (1.47)$$

to rotation,

$$\gamma_{0i} = \left(-\frac{\Omega y}{c}, \frac{\Omega x}{c}, 0\right) s$$

one gets

$$\xi = -\frac{1}{2}\int dz^0 \gamma_{i0,j}(z) S^{ij} = \int dt \Omega S_z,$$

where $S_z = S^{12}$. In general, one can write

$$\xi = \int dt \boldsymbol{\Omega} \cdot \mathbf{S},$$

which must now be applied to a field-free solution $a_\mu(x)$. One again finds

$$E'_\pm = E + \boldsymbol{\Omega} \cdot \mathbf{S}$$

and, for photons polarized parallel or antiparallel to $\boldsymbol{\Omega}$,

$$E'_\pm = E \pm \hbar\Omega,$$

as in Ref. [18]. E' is the energy observed by the co-rotating observer. It must be multiplied by a factor γ when referred to a non-rotating observer.

The solution found is exact to first order in $\gamma_{\mu\nu}$. It also is Lorentz invariant and invariant under gravitational gauge transformations [38] and provides, to the order considered, the general relativistic generalization of the notion of wave vector. Such a generalization appears intrinsically nonlocal, as indicated by the presence of the angular momentum operator in the expressions for ξ. Unlike the standard approach, the results apply to both inertial and gravitational fields and do not depend on the eikonal approximation. In addition, the gravity-particle coupling becomes effectively vectorial, as previously found for the spin-0 case [40,55]. When $\gamma_{\mu\nu}$ represents rotation, the results reproduce those of Mashhoon and Hehl and Ni.

1.3.4 The Dirac Equation in Curved Space–Time

Some fundamental properties of the vierbein formalism must be recalled. The formalism is required to study the Dirac equation in curved space–time.

In passing from flat to curved space–time, the standard prescriptions $\partial \to \nabla$, and $\eta_{\mu\nu} \to g_{\mu\nu}$ suggested by the equivalence principle, are used. The procedure to replace flat space tensors with "curved space" tensor cannot, however, be extended to the case of spinors. This procedure works with tensors because the tensor representations of $GL(4, R)$, the group of 4×4 real matrices, behave like tensors under the subgroup $SO(3, 1)$. Thus, considering the vector representation, as an example, one gets

$$V'^\mu(x') = \frac{\partial x'^\mu}{\partial x^\nu} V^\nu(x) \quad \overset{x'=\Lambda^\mu_{\ \nu} x^\nu}{\longleftrightarrow} \quad V'^\mu(x') = \Lambda^\mu_{\ \nu} V^\nu(x).$$

But there are no representations of $GL(4, R)$ which behave like spinors under $SO(3, 1)$, i.e. there does not exist a function of x and x' which reduces to the usual spinor representation of the Lorentz group ($D(\Lambda)$) for $x' = \Lambda^{\mu}{}_{\nu} x^{\nu}$. In order to write the general covariant coupling of spin-1/2 particles to gravity one must use the vierbein formalism.

A vierbein (or tetrad) field is defined as

$$e^{a}{}_{\mu}(X) = \frac{\partial \xi^{a}(x)}{\partial x^{\mu}}|_{x=X},$$

where ξ^a represents the local inertial coordinates, x^μ the generic coordinate, $\xi^a(x) \to \xi'^a(x) = \Lambda^a_b(x)\xi^b(x)$, and $\Lambda^T \eta \Lambda = \eta$.

The quantities $e^a{}_\mu(x)$ constitute a set of four coordinate vectors which form a basis for the (flat) tangent space to the curved space at the point $x = X$. Under the coordinate transformation $x \to x' = x'(x)$, the vierbeins $e^a{}_\mu(x)$ transform as

$$e'^{a}{}_{\mu}(x') = \frac{\partial x^{\nu}}{\partial x'^{\mu}} e^{a}{}_{\nu}(x), \quad (\xi'^{a}(x') = \xi^{a}(x)).$$

The metric $g_{\mu\nu}(x)$ is related to the vierbein fields by the relation $g_{\mu\nu}(x) = \eta_{ab}\, e^a{}_\mu(x) e^b{}_\nu(x)$, where η_{ab} is the Minkowski metric (in the local inertial frame). It follows that

$$\delta^{\mu}_{\nu} = e_a{}^{\mu}(x) e^{a}{}_{\nu}(x) \quad \text{i.e. } e^{a}{}_{\nu}(x) \text{ is the inverse of } e_a{}^{\nu}(x)$$

$$\Downarrow$$

$$\eta^{ab} = g^{\mu\nu}(x) e^{a}{}_{\mu}(x) e^{b}{}_{\nu}(x).$$

Spinor fields are coordinate scalars that transform, under local Lorentz transformations, as

$$\psi_\alpha(x) \to \psi'_\alpha(x) = D_{\alpha\beta}[\Lambda(x)]\psi_\beta(x),$$

where $D_{\alpha\beta}[\Lambda(x)]$ is the spinor representation of the Lorentz group and ψ_α is the component of the spinor ψ (not to be confused with general coordinate indices). Since Λ *does depend* on x, $\partial_\mu \psi_\alpha$ *does not* transform like ψ_α under local Lorentz transformations. To obtain a Lagrangian invariant under generic coordinate transformations one must define the covariant derivative

$$D_\mu \psi_\alpha \equiv \partial_\mu \psi_\alpha - [\Omega_\mu]_{\alpha\beta}\psi_\beta,$$

where $[\Omega_\mu]_{\alpha\beta}$ is the connection matrix. The appropriate requirements therefore are

$$D_\mu \psi_\alpha \to D_{\alpha\beta}[\Lambda(x)]\, D_\mu \psi_\beta(x)$$

and

$$\Omega'_\mu = D(\Lambda)\Omega_\mu D^{-1}(\Lambda) - (\partial_\mu D(\Lambda))^{-1} D^{-1}(\Lambda)$$

The connection matrix $[\Omega_\mu]_{\alpha\beta}(x)$ can be written as

$$[\Omega_\mu]_{\alpha\beta}(x) = \frac{i}{2}[S_{ab}]_{\alpha\beta}\omega_\mu{}^{ab}(x),$$

where $S_{ab} = \frac{\sigma_{ab}}{2} = \frac{i[\gamma_a,\gamma_b]}{2}$ are the generators of the Lorentz group in the spinor representation and $\omega_\mu{}^a{}_b$ represent the spin connections. The spinor representation of the Lorentz group can be written as $D[\Lambda(x)] = \exp[-(i/2)S_{ab}\theta^{ab}]$. The covariant derivative acts on vierbeins according to

$$D_\mu e^a{}_\nu = \partial_\mu e^a{}_\nu - \Gamma^\lambda_{\mu\nu} e^a{}_\lambda - \omega_\mu{}^a{}_b e^b{}_\nu,$$

where the $\Gamma^\mu_{\alpha\beta}$ are the usual Christoffel symbols and the condition $D_\mu e^a{}_\nu = 0$ can be used to determine the spin connection coefficients, also known as Ricci rotation coefficients,

$$\omega_{bca} = e_{b\lambda}\left(\partial_a e^\lambda{}_c + \Gamma^\lambda_{\gamma\mu} e^\gamma_c e^\mu_a\right).$$

The general covariant coupling of spin-1/2 particles to gravity is therefore given by the Lagrangian

$$\mathcal{L} = \sqrt{-g}(\bar\psi\gamma^a D_a\psi - m\bar\psi\psi), \tag{1.48}$$

where $D_a = \partial_a - \frac{i}{4}\omega_{bca}\sigma^{bc}$ is the covariant derivative introduced above. The Lagrangian is invariant under the local Lorentz transformation of the vierbein and the spinor fields.

By using the properties of the Dirac matrices

$$\gamma^a[\gamma^b,\gamma^c] = \eta^{ab}\gamma^c + \eta^{ac}\gamma^b - i\varepsilon^{dabc}\gamma_d\gamma^5 \tag{1.49}$$

the Lagrangian density (1.48) can be written in the form

$$\mathcal{L} = \det(e)\,\bar\psi\left(i\gamma^a\partial_a - m - \gamma_5\gamma_d B^d\right)\psi, \tag{1.50}$$

where

$$B^d = \epsilon^{abcd} e_{b\lambda}(\partial_a e^\lambda_c + \Gamma^\lambda_{\alpha\mu} e^\alpha_c e^\mu_a).$$

In a local inertial frame of the fermion, the effect of a gravitational field appears as the *axial-vector interaction* term shown in $\mathcal{L}$.

The Dirac equation in curved space–time is given by [61]

$$\left(i\gamma^\mu(x)\mathcal{D}_\mu - m\right)\psi = 0, \tag{1.51}$$

where the covariant derivative $\mathcal{D}_\mu$ is defined as

$$\mathcal{D}_\mu = \partial_\mu + \Gamma_\mu(x), \tag{1.52}$$

and $\Gamma_\mu(x)$ is the spin connection

$$\Gamma_\mu(x) = -\frac{i}{4}\omega_{\hat\beta\hat\gamma\mu}\sigma^{\hat\beta\hat\gamma}\,. \tag{1.53}$$

In this equation

$$\omega_{\hat\beta\hat\gamma\mu} = e_{\hat\beta\lambda}e_{\hat\gamma}{}^{\lambda}{}_{;\mu} = e_{\hat\beta\lambda}\left(\partial_\mu e_{\hat\gamma}{}^{\lambda} + \Gamma^{\lambda}_{\nu\mu}e_{\hat\gamma}{}^{\nu}\right)\,. \tag{1.54}$$

The matrices $\gamma^\mu(x)$ are related to the usual Dirac matrices $\gamma^{\hat\alpha}$ by means of the vierbein fields $e^{\mu}_{\hat\alpha}(x)$, i.e. $\gamma^\mu(x) = e^{\mu}_{\hat\alpha}(x)\gamma^{\hat\alpha}$. The indices with and without hats refer to flat and curved space–time, respectively. The equations derived are general and completely independent of the representation of the usual Dirac matrices that must be specified in the applications. The vierbein vector fields satisfy the relations

$$\eta^{\hat\alpha\hat\beta}e_{\hat\alpha}{}^{\mu}e_{\hat\beta}{}^{\nu} = g^{\mu\nu}\,, \quad g_{\mu\nu}e_{\hat\alpha}{}^{\mu}e_{\hat\beta}{}^{\nu} = \eta_{\hat\alpha\hat\beta}\,, \tag{1.55}$$

where $\eta_{\hat\alpha\hat\beta} = \mathrm{diag}\,(1, -1, -1, -1)$ is the Minkowski metric of flat space–time and $g_{\mu\nu}(x)$ is the curved space–time metric.

The spin connection coefficients (1.54) can be rewritten as

$$\begin{aligned}\omega_{\hat\beta\hat\gamma\mu} = \frac{1}{2}\Big[& e_{\hat\beta}{}^{\lambda}(\partial_\mu e_{\hat\gamma\lambda} - \partial_\lambda e_{\hat\gamma\mu}) + e_{\hat\gamma}{}^{\lambda}(\partial_\lambda e_{\hat\beta\mu} - \partial_\mu e_{\hat\beta\lambda}) + \\ & + e_{\hat\beta}{}^{\lambda}e_{\hat\gamma}{}^{\nu}e_{\hat\sigma\mu}(\partial_\nu e^{\hat\sigma}{}_{\lambda} - \partial_\lambda e^{\hat\sigma}{}_{\nu})\Big]\,,\end{aligned} \tag{1.56}$$

With this construction, the density Lagrangian is invariant under general coordinate transformations and local Lorentz transformations.

The linear theory of gravity is obtained by writing the metric tensor in the form

$$g_{\mu\nu} = \eta_{\mu\nu} + \gamma_{\mu\nu}\,, \tag{1.57}$$

where the tensor $\gamma_{\mu\nu}$ ($\ll \eta_{\mu\nu}$) represents a perturbation, identified with the graviton in the language of the quantum theory of fields. For the vierbein fields one then finds

$$e_{\hat\alpha\mu} = \eta_{\alpha\mu} + \gamma_{\hat\alpha\mu}\,, \quad e_{\hat\alpha}{}^{\mu} = \delta^{\mu}_{\alpha} - \frac{1}{2}\gamma_{\hat\alpha}{}^{\mu}\,, \quad \det(e) = 1 + \gamma\,, \quad \gamma = \eta_{\mu\nu}\gamma^{\mu\nu}\,. \tag{1.58}$$

In (1.58) and hereafter, the "hat" symbol on indices is removed whenever confusion does not arise, and the general and local coordinates coincide.

1.3.5 The Lense–Thirring Metric

The procedure outlined in the previous section can be immediately applied to the line element of a rotating gravitational source [62] that, in WFA, is simply described by [61]

$$ds^2 = (1-\phi)(dt)^2 - (1+\phi)(d\mathbf{x})^2 - 2\boldsymbol{\gamma}\cdot d\mathbf{x}dt\,, \tag{1.59}$$

where

$$\mathbf{x} = (x, y, z)\,, \quad \phi = \frac{2GM}{r}\,, \quad \boldsymbol{\gamma} = \frac{4GMR^2}{5r^3}\boldsymbol{\omega}\times\mathbf{x}\,,$$

M is the mass of the gravitational source and R its radius. The corresponding vierbein fields are [63]

$$e_{\hat{0}0} = (1+\phi/2)\,, \quad e_{\hat{i}j} = -(1-\phi/2)\,\delta_{ij}\,, \quad e_{\hat{i}0} = \gamma_{\hat{i}0} = -\gamma^i\,, \tag{1.60}$$

and all the other components are zero. The off-diagonal terms in (1.60) make the vierbein fields asymmetric. One also finds

$$\begin{aligned}\Gamma_0 &= -\frac{1}{2}\varphi,_j\,\sigma^{\hat{0}\hat{j}} - \frac{1}{4}(\gamma_{i,j}-\gamma_{j,i})\sigma^{\hat{i}\hat{j}} \\ \Gamma_i &= -\frac{1}{4}(\gamma_{i,j}-\gamma_{j,i})\sigma^{\hat{0}\hat{j}} - \frac{1}{2}\varphi,_j\,\sigma^{\hat{i}\hat{j}}\,.\end{aligned} \tag{1.61}$$

For geometries with spherical symmetry, such as the Schwarzschild space–time, γ^i vanishes and from (1.60) one gets

$$\gamma_{\hat{0}0} = \frac{\phi}{2} = \gamma_{00}\,, \quad \gamma_{\hat{i}j} = \frac{\phi}{2}\,\delta_{ij} = \gamma_{ji}\,, \tag{1.62}$$

that is the vierbein fields are symmetric. In the linear approximation, therefore, it is no longer necessary to distinguish between general and local indices.

Equations (1.60) and (1.61) will be used repeatedly in Chaps. 2 and 3.

Tetrads for accelerating and rotating systems

Consider the line element $ds^2 = g_{\mu\nu}(x)dx^\mu dx^\nu$ and the set of tangent vectors $\mathbf{e}_\mu = \partial_\mu\mathbf{P}$ that forms the coordinate basis spanning the manifold $g_{\mu\nu}(x) = \partial_\mu\mathbf{P}\cdot\partial_\nu\mathbf{P} \equiv \mathbf{e}_\mu\cdot\mathbf{e}_\nu$. The principle of equivalence ensures the existence of an orthonormal tetrad frame $\mathbf{e}_{\hat{\mu}} = \partial_{\hat{\mu}}\mathbf{P}$ such that for a local tangent space defined at any given point of space–time one has $\eta_{\hat{\mu}\hat{\nu}} = \mathbf{e}_{\hat{\mu}}\cdot\mathbf{e}_{\hat{\nu}}$. The principle underlying the tetrad formalism therefore requires that, for a sufficiently small region of space–time, $\mathbf{e}_\mu$ be mapped onto $\mathbf{e}_{\hat{\mu}}$ using a set of projection functions $e^{\nu}_{\hat{\mu}}$ and their inverses $e^{\nu}_{\hat{\mu}}$ such that (see (1.55))

$$\begin{aligned} e_{\hat{\mu}} &= e^{\nu}_{\hat{\mu}}\, e_{\nu}, \quad e_{\mu} = e^{\hat{\nu}}_{\mu}, \quad \eta_{\hat{\mu}\hat{\nu}} = e^{\mu}_{\hat{\mu}} e^{\nu}_{\hat{\nu}} g_{\mu\nu}(x), \\ e^{\nu}_{\hat{\mu}} e^{\hat{\alpha}}_{\nu} &= \delta^{\hat{\alpha}}_{\hat{\mu}}, \, e^{\hat{\nu}}_{\mu} e^{\alpha}_{\hat{\nu}} = \delta^{\alpha}_{\mu}. \end{aligned} \tag{1.63}$$

When e_{μ} refers to an observer with acceleration $\mathbf{a}$ rotating with angular velocity $\boldsymbol{\omega}$, one finds [14]

$$ds^2 = [(1+\mathbf{a}\cdot\mathbf{x})^2 + (\boldsymbol{\omega}\cdot\mathbf{x})^2 - \omega^2 x^2]dx_0^2 - 2dx_0 d\mathbf{x}\cdot(\boldsymbol{\omega}\times\mathbf{x}) - d\mathbf{x}\cdot d\mathbf{x}\,, \tag{1.64}$$

while

$$\begin{aligned} e_{\hat{0}} &= (1+\mathbf{a}\cdot\mathbf{x})^{-1}[\mathbf{e}_0 - (\boldsymbol{\omega}\times\mathbf{x})^k e_k], \qquad e_{\hat{i}} = e_i\,, \\ e^0_{\hat{0}} &= (1+\mathbf{a}\cdot\mathbf{x})^{-1}, \qquad e^k_{\hat{0}} = -(1+\mathbf{a}\cdot\mathbf{x})^{-1}\boldsymbol{\omega}\times\mathbf{x}^k), \\ e^{\hat{0}}_0 &= 1+\mathbf{a}\cdot\mathbf{x},\, e^{\hat{k}}_0 = (\boldsymbol{\omega}\times\mathbf{x})^k, \qquad e^{\hat{0}}_i = 0, \quad e^{\hat{k}}_i = \delta^k_i\,. \end{aligned} \tag{1.65}$$

Also, from $D_{\mu}\gamma_{\nu}(x) = 0$ and $\gamma_{\mu}(x) = e^{\hat{\mu}}_{\mu}\gamma_{\hat{\mu}}$, where $\gamma_{\hat{\mu}}$ represents the usual Dirac matrices, one finds $\Gamma_{\mu}(x) = \frac{1}{4}\sigma^{\alpha\hat{\beta}}\omega_{\alpha\hat{\beta}\mu}e^{\hat{\mu}}_{\mu}$. The Ricci coefficients are

$$\Gamma_{\nu\hat{\alpha}\beta} = \frac{1}{2}(C_{\nu\hat{\alpha}\beta} + C_{\alpha\hat{\beta}\nu} - C_{\beta\hat{\nu}\alpha})\,,$$

and

$$C_{\nu\hat{\alpha}\beta} = \eta_{\mu\nu}e^{\alpha}_{\hat{\alpha}}e^{\beta}_{\hat{\beta}}(\partial_{\alpha}e^{\hat{\mu}}_{\beta} - \partial^{\hat{\mu}}_{\beta})\,.$$

It also follows that

$$\Gamma_0 = -\frac{1}{2}a_i\sigma^{0i} - \frac{1}{2}\boldsymbol{\omega}\cdot\boldsymbol{\sigma}, \Gamma_i = 0\,,$$

with $\sigma^{0i} = \frac{1}{2}[\gamma^0, \gamma^i]$.

The generally covariant Dirac equation

Within the context of general relativity, De Oliveira and Tiomno [64] and Peres [65] have conducted comprehensive studies of the fully covariant Dirac equation. The latter takes the form

$$[i\gamma^{\mu}(x)\mathcal{D}_{\mu} - m]\Psi(x) = 0, \tag{1.66}$$

where $\mathcal{D}_{\mu} = \nabla_{\mu} + i\Gamma_{\mu}(x)$, ∇_{μ} is the covariant derivative, $\Gamma_{\mu}(x)$ the spin connection and the matrices $\gamma(x)$ satisfy the relations $\{\gamma^{\mu}(x), \gamma^{\nu}(x)\} = 2g^{\mu\nu}$. The relations connecting $\Gamma_{\mu}(x)$ and $\gamma^{\mu}(x)$ to the tetrad are given in Sects. 1.3.4 and 1.3.5. Equation

(1.66) can be solved exactly to first order in $\gamma_{\mu\nu}(x)$ by first transforming it into the equation

$$[i\tilde{\gamma}^{\nu}(x)\nabla_{\nu} - m]\tilde{\Psi}(x) = 0\,, \tag{1.67}$$

where

$$\tilde{\Psi}(x) = S^{-1}\Psi(x)\,, \qquad S(x) = e^{-i\Phi_s(x)}\,, \qquad \Phi_s(x) = \mathcal{P}\int_P^x dz^{\lambda}\Gamma_{\lambda}(z)\,, \tag{1.68}$$

$$\tilde{\gamma}(x) = S^{-1}\gamma(x)S\,.$$

The presence of redundant solutions which do not satisfy the initial first-order equation is excluded by multiplying (1.67) on the left by $(-i\tilde{\gamma}^{\nu}(x)\nabla_{\nu} - m)$. One obtains the equation

$$(g^{\mu\nu}\nabla_{\mu}\nabla_{\nu} + m^2)\tilde{\Psi}(x) = 0\,, \tag{1.69}$$

which has the first-order exact solution

$$\tilde{\Psi}(x) = e^{-i\hat{\Phi}_G(x)}\Psi_0(x)\,, \tag{1.70}$$

already discussed in connection with the Klein–Gordon equation. The operators $\hat{\Phi}_G(x)$, $\hat{L}^{\alpha\beta}$ and $\hat{k}^{\alpha}$ are defined by

$$\hat{\Phi}_G(x) = -\frac{1}{4}\int_P^x dz^{\lambda}\left[\gamma_{\alpha\lambda,\beta}(z) - \gamma_{\beta\lambda,\alpha}(z)\right]\hat{L}^{\alpha\beta}(z) + \frac{1}{2}\int_P^x dz^{\lambda}\gamma_{\alpha\lambda}\hat{k}^{\alpha}\,, \tag{1.71}$$

$$[\hat{L}^{\alpha\beta}(z), \Psi_0(x)] = \left((x^{\alpha} - z^{\alpha})\hat{k}^{\beta} - (x^{\beta} - z^{\beta})\hat{k}^{\alpha}\right)\Psi_0(x)\,,$$

$$[\hat{k}^{\alpha}, \Psi_0(x)] = i\partial^{\alpha}\Psi_0\,,$$

and $\Psi_0(x)$ satisfies the usual flat space–time Dirac equation. $L_{\alpha\beta}$ and k^{α} are the angular and linear momenta of the free particle, respectively. It follows from (1.67) and (1.68) that the solution of (1.66) can be written in the form

$$\Psi(x) = e^{-i\Phi_s}\left(-i\tilde{\gamma}^{\mu}(x)\nabla_{\mu} - m\right)e^{-i\Phi_G}\,\Psi_0(x)\,, \tag{1.72}$$

and also as

$$\Psi(x) = -\frac{1}{2m}\left(-i\gamma^{\mu}(x)\mathcal{D}_{\mu} - m\right)e^{-i\Phi_T}\,\Psi_0(x) \equiv \hat{T}\Psi_0\,, \tag{1.73}$$

where $\Phi_T = \Phi_s + \Phi_G$, as well as $\gamma^{\mu}(x)$, are first order quantities in $\gamma_{\alpha\beta}(x)$. By multiplying (1.66) on the left by $(-i\gamma^{\nu}(x)\mathcal{D}_{\nu} - m)$ and using the relations

$$\nabla_{\mu}\Gamma_{\nu}(x) - \nabla_{\nu}\Gamma_{\mu}(x) + i[\Gamma_{\mu}(x), \Gamma_{\nu}(x)] = -\frac{1}{4}\sigma^{\alpha\beta}(x)R_{\alpha\beta\mu\nu}\,, \tag{1.74}$$

and

$$[\mathcal{D}_\mu, \mathcal{D}_\nu] = -\frac{i}{4}\sigma^{\alpha\beta}(x)R_{\alpha\beta\mu\nu}\,, \tag{1.75}$$

one obtains the equation

$$\left(g^{\mu\nu}\mathcal{D}_\mu\mathcal{D}_\nu - \frac{R}{4} + m^2\right)\Psi(x) = 0\,. \tag{1.76}$$

In (1.75) and (1.76) $R_{\alpha\beta\mu\nu}$ is the (linearized) Riemann tensor, and R the corresponding Ricci scalar. In the transformation from first order to second-order wave equation curvature appears and with it an effective mass $\tilde{m} = \sqrt{m^2 - R/4}$, as first pointed out by Weyl [66]. Notice that $\tilde{m}$ does not vanish when $m = 0$ and that $R \neq 0$ in its linearized form.

By using Eq. (1.68), one also finds

$$\left(-i\gamma^\nu(x)\mathcal{D}_\nu - m\right) S\left(i\tilde{\gamma}^\mu\nabla_\mu - m\right)\tilde{\Psi}(x) = S\left(g^{\mu\nu}\nabla_\mu\nabla_\nu + m^2\right)\tilde{\Psi}(x) = 0\,. \tag{1.77}$$

On applying Stokes theorem to a closed space–time path C, one obtains from Φ_T and (1.74)

$$\Delta\Phi_T = \frac{1}{4}\int_\Sigma d\tau^{\mu\nu} J^{\alpha\beta} R_{\mu\nu\alpha\beta}\,, \tag{1.78}$$

where Σ is a surface bound by C and $J^{\alpha\beta} = L^{\alpha\beta} + \sigma^{\alpha\beta}$ is the total angular momentum of the particle. Equation (1.78) confirms that the gyro-gravitational ratio of a Dirac particle is 1.

Choose a cylindrical coordinate (t, r, θ, z) for an inertial frame F_0. An observer at rest in a frame F' rotating with a constant angular velocity Ω relative to F_0 will follow the world line

$$r = \text{const.}, \theta = \text{const.} + \Omega t, z = \text{const.}$$

A useful orthogonal tetrad consists of the observer's four-velocity $\lambda^\mu_{(0)} = dx^\mu/ds$ and the triad $\lambda^\mu_{(i)}$ $(i = 1,2,3)$ normal to the world line. Using the local tetrad [67,68]

$$\lambda^\mu_{(0)} = \left(\gamma, 0, \frac{\gamma\Omega}{c}, 0\right), \qquad \lambda^\mu_{(1)} = (0, 1, 0, 0)\,, \tag{1.79}$$

$$\lambda^\mu_{(2)} = \left(\frac{\gamma\Omega r}{c}, 0, \frac{\gamma}{r}, 0\right), \qquad \lambda^\mu_{(3)} = (0, 0, 0, 1),$$

where $\gamma \equiv (1 - r^2\Omega^2/c^2)^{-1/2}$, one can construct [68] a vierbein field $e^\mu{}_a(x)$ along the world line of the observer

$$e^{\mu}{}_{(0)} = \left(\gamma, 0, \frac{\gamma\Omega}{c}, 0\right) \qquad e^{\mu}{}_{(1)} = \left(-\frac{\gamma\Omega r}{c}\sin\gamma\Omega t, \cos\gamma\Omega t, \frac{\gamma}{r}\sin\gamma\Omega t, 0\right) \tag{1.80}$$

$$e^{\mu}{}_{(2)} = \left(\frac{\gamma\Omega r}{c}\cos\gamma\Omega t, \sin\gamma\Omega t, \frac{\gamma}{r}\cos\gamma\Omega t, 0\right) \qquad e^{\mu}{}_{(3)} = (0, 0, 0, 1). \tag{1.81}$$

It is then easy to calculate the spinor connection Γ_{μ}. In calculating the energy, only the component Γ_0 is necessary. By using the Dirac representation [69] for the γ matrices (Appendix B), one obtains $\Gamma_0 = \frac{\gamma\Omega}{2c}\sigma_z$ and from Eq. (1.2)

$$e^{-i\int \Gamma_0 dz^0}\,\Psi = e^{-\frac{1}{2}\int \gamma\Omega\sigma_z dt}\,\Psi,$$

where Ψ has the usual plane wave form. Besides the contribution due to the coupling of the orbital angular momentum to rotation, which gives the Sagnac effect [38,70], one obtains the spin-rotation coupling

$$E' = E + \frac{\hbar}{2}\Omega\sigma_z.$$

For spin polarizations parallel or antiparallel to the direction of rotation one obtains,

$$E'_{\pm} = E \pm \frac{\hbar}{2}\Omega,$$

as in Ref. [18]. The present result is exact and follows as a special case from the general coupling of Eq. (1.2). It also agrees with that of Hehl and Ni [14] who have derived a variety of other inertial effects for fermions, by transforming the special-relativistic Dirac equation to a non-inertial frame.

1.3.6 Spin-2 Particles in Gravitational Fields

The study of wave equations in the WFA can be extended to massive as well as massless spin-2 particles. The covariant Fierz–Pauli equation is

$$\nabla_{\alpha}\nabla^{\alpha}\Phi_{\mu\nu} + m^2\Phi_{\mu\nu} = 0, \tag{1.82}$$

where m is the mass of the particle. The solution of (1.82) is expected to yield quantum phases as in the previous instances.

The choice of (1.82) as the proper covariant equation to be studied, requires some justification. The propagation equations of higher spin fields contain in general curvature dependent terms that make the formulation of these fields particularly difficult when $m = 0$ [71]. For spin-2 fields, the simplest equation of propagation used in lensing is derived in [72] and is given by

$$\nabla_{\alpha}\nabla^{\alpha}\Phi_{\mu\nu} + 2R_{\alpha\mu\beta\nu}\Phi^{\alpha\beta} = 0. \tag{1.83}$$

The second term in (1.83) is localized in a region surrounding the lens that is small relative to the distances between lens, source and observer and is neglected when the wavelength λ, associated with $\Phi_{\mu\nu}$, is smaller than the typical radius of curvature of the gravitational background [73]. For the metrics used below, one finds, in particular, that the curvature term may be neglected when $\lambda \ll \sqrt{\frac{\rho^3}{r_g}}$, where ρ is the distance from the lens and r_g its Schwarzschild radius. This condition is satisfied in most lensing problems. It also is adequate to treat the problems discussed in [74,75].

A mass term can also be added to Eq. (1.83)

$$\nabla_\alpha \nabla^\alpha \Phi_{\mu\nu} + 2R_{\alpha\mu\beta\nu}\Phi^{\alpha\beta} + m^2\Phi_{\mu\nu} = 0\,. \tag{1.84}$$

Here too the curvature term is smaller than the mass term whenever $m > 1/r_g$. For Earth-bound experiments $r_g = 2GM_\oplus/R_\oplus$ and the curvature term becomes negligible for $m > 2.5 \cdot 10^{-6} GeV$. In view of the applications discussed below, the curvature term is negligible and (1.84) reduces to the initial Eq. (1.82).

To first order in $\gamma_{\mu\nu}$, (1.82) can be written in the form

$$\left(\eta^{\alpha\beta} - \gamma^{\alpha\beta}\right)\partial_\alpha\partial_\beta\Phi_{\mu\nu} + R_{\sigma\mu}\Phi^\sigma_\nu + R_{\sigma\nu}\Phi^\sigma_\mu - 2\Gamma^\sigma_{\mu\alpha}\partial^\alpha\Phi_{\nu\sigma} - \tag{1.85}$$

$$-2\Gamma^\sigma_{\nu\alpha}\partial^\alpha\Phi_{\mu\sigma} + m^2\Phi_{\mu\nu} = 0,$$

where $R_{\mu\beta} = -(\frac{1}{2})\partial_\alpha\partial^\alpha\gamma_{\mu\beta}$ is the linearized Ricci tensor of the background metric and $\Gamma_{\sigma\mu,\alpha} = \frac{1}{2}\left(\gamma_{\alpha\sigma,\mu} + \gamma_{\alpha\mu,\sigma} - \gamma_{\sigma\mu,\alpha}\right)$ are the corresponding Christoffel symbols of the first kind.

Solution of the spin-2 wave equation

It is easy to prove, by direct substitution, that a solution of (1.85), exact to first order in $\gamma_{\mu\nu}$, is represented by

$$\Phi_{\mu\nu} = \phi_{\mu\nu} - \frac{1}{4}I_1 + \frac{1}{2}I_2 \tag{1.86}$$

$$I_1 \equiv \int_P^x dz^\lambda \left(\gamma_{\alpha\lambda,\beta}(z) - \gamma_{\beta\lambda,\alpha}(z)\right)\left[\left(x^\alpha - z^\alpha\right)\partial^\beta\phi_{\mu\nu}(x) - \left(x^\beta - z^\beta\right)\partial^\alpha\phi_{\mu\nu}(x)\right],$$

$$I_2 \equiv \int_P^x dz^\lambda\gamma_{\alpha\lambda}(z)\,\partial^\alpha\phi_{\mu\nu}(x) + \int_P^x dz^\lambda\Gamma_{\mu\lambda,\sigma}(z)\,\phi^\sigma_\nu(x) + \int_P^x dz^\lambda\Gamma_{\nu\lambda,\sigma}(z)\,\phi^\sigma_\mu(x),$$

where $\phi_{\mu\nu}$ satisfies the field-free equation

$$\left(\partial_\alpha\partial^\alpha + m^2\right)\phi_{\mu\nu}(x) = 0\,. \tag{1.87}$$

The Lanczos–DeDonder gauge condition has also been used.

The particular case $m = 0$ yields the solution $\Phi_{\mu\nu}$ for a linearized gravitational field $\phi_{\mu\nu}$ propagating in a background gravitational field $\gamma_{\mu\nu}$.

The solution (1.86) applies equally well when $\phi_{\mu\nu}$ is a plane wave or a wave packet solution of (1.87). No additional approximations are made regarding $\Phi_{\mu\nu}$ that obviously satisfies Eq. (1.87) when $\gamma_{\mu\nu}$ vanishes.

It is shown below that, as in [55,56], the solution is manifestly covariant. It also is completely gauge invariant and the effect of gravitation is entirely contained in the phase of the wave function. In fact, Eq. (1.86) can be written in the form $\Phi_{\mu\nu} = exp(-i\xi)\phi_{\mu\nu} \simeq (1 - i\xi)\phi_{\mu\nu}$ or, explicitly, as

$$\Phi_{\mu\nu}(x) = \phi_{\mu\nu}(x) + \frac{1}{2}\int_P^x dz^\lambda \gamma_{\alpha\lambda}(z)\,\partial^\alpha \phi_{\mu\nu}(x) - \tag{1.88}$$
$$-\frac{1}{2}\int_P^x dz^\lambda \left(\gamma_{\alpha\lambda,\beta}(z) - \gamma_{\beta\lambda,\alpha}(z)\right)\left[\left(x^\alpha - z^\alpha\right)\partial^\beta + iS^{\alpha\beta}\right]\phi_{\mu\nu}(x) -$$
$$-\frac{i}{2}\int_P^x dz^\lambda \gamma_{\beta\sigma,\lambda}(z)\,T^{\beta\sigma}\phi_{\mu\nu}(x)\,,$$

where

$$S^{\alpha\beta}\phi_{\mu\nu} \equiv \frac{i}{2}\left(\delta^\alpha_\sigma\delta^\beta_\mu\delta^\tau_\nu - \delta^\beta_\sigma\delta^\alpha_\mu\delta^\tau_\nu + \delta^\alpha_\sigma\delta^\beta_\nu\delta^\tau_\mu - \delta^\beta_\sigma\delta^\alpha_\nu\delta^\tau_\mu\right)\phi^\sigma_\tau \tag{1.89}$$
$$T^{\beta\sigma}\phi_{\mu\nu} \equiv i\left(\delta^\beta_\mu\delta^\tau_\nu + \delta^\beta_\nu\delta^\tau_\mu\right)\phi^\sigma_\tau\,.$$

From $S^{\alpha\beta}$ one constructs the rotation matrices $S_i = -2i\epsilon_{ijk}S^{jk}$ that satisfy the commutation relations $[S_i, S_j] = i\epsilon_{ijk}S_k$. The spin-gravity interaction is therefore contained in the term

$$\Phi'_{\mu\nu} \equiv -\frac{i}{2}\int_P^x dz^\lambda \left(\gamma_{\alpha\lambda,\beta} - \gamma_{\beta\lambda,\alpha}\right)S^{\alpha\beta}\phi_{\mu\nu}(x) = \tag{1.90}$$
$$= \frac{1}{2}\int_P^x dz^\lambda \left[\left(\gamma_{\sigma\lambda,\mu} - \gamma_{\mu\lambda,\sigma}\right)\phi^\sigma_\nu + \left(\gamma_{\sigma\lambda,\nu} - \gamma_{\nu\lambda,\sigma}\right)\phi^\sigma_\mu\right].$$

The solution (1.86) is invariant under the gauge transformations $\gamma_{\mu\nu} \rightarrow \gamma_{\mu\nu} - \xi_{\mu,\nu} - \xi_{\nu,\mu}$, where ξ_μ are small quantities of the first order. On choosing a closed integration path Γ, Stokes theorem transforms the first three integrals of (1.88) into $1/4\int_\Sigma d\sigma^{\lambda\kappa}R_{\lambda\kappa\alpha\beta}\left(L^{\alpha\beta} + S^{\alpha\beta}\right)\phi_{\mu\nu}$, where Σ is the surface bounded by Γ, $J^{\alpha\beta} = L^{\alpha\beta} + S^{\alpha\beta}$ is the total angular momentum of the particle and $R_{\lambda\kappa\alpha\beta}$ is the linearized Riemann tensor. For the same path Γ the integral involving $T^{\beta\sigma}$ in (1.88) vanishes. It behaves like a gauge term and may therefore be dropped. For the same closed paths, (1.88) gives

$$\Phi_{\mu\nu} \simeq (1 - i\xi)\phi_{\mu\nu} = \left(1 - \frac{i}{4}\int_\Sigma d\sigma^{\lambda\kappa}R_{\lambda\kappa\alpha\beta}J^{\alpha\beta}\right)\phi_{\mu\nu}\,, \tag{1.91}$$

which obviously is covariant and gauge invariant. For practical applications, Eq. (1.86) is easier to use.

Helicity-gravity coupling and geometrical optics

The helicity-rotation coupling for massless, or massive spin-2 particles follows immediately from the $S^{\alpha\beta}$ term in Eq. (1.88). In fact, the particle energy is changed by virtue of its spin by an amount given by the time integral of this spin term

$$\xi^{hr} = -\frac{1}{2}\int_P^x dz^0\left(\gamma_{\alpha 0,\beta} - \gamma_{\beta 0,\alpha}\right)S^{\alpha\beta}\,, \tag{1.92}$$

that must then be applied to a solution of (1.87). For rotation about the x^3-axis, $\gamma_{0i} = \Omega(y, -x, 0)$, we find $\xi^{hr} = -\int_P^x dz^0 2\Omega S^3$ and the energy of the particle therefore changes by $\pm 2\Omega$, where the factor ± 2 refers to the particle's helicity, as discussed by Ramos and Mashhoon [75]. Equation (1.92) extends their result to any weak gravitational, or inertial field.

The effect of (1.90) on $\phi_{\mu\nu}$ can be easily seen in the case of a gravitational-wave propagating in the x-direction and represented by the components

$$\phi_{22} = -\phi_{33} = \varepsilon_{22}e^{ik(t-x)}$$

and

$$\phi_{23} = \varepsilon_{23}e^{ik(t-x)}\,.$$

For an observer rotating about the x-axis the metric is $\gamma_{00} = -\Omega^2 r^2\,, \gamma_{11} = \gamma_{22} = \gamma_{33} = -1\,, \gamma_{0i} = \Omega(0, z, -y)$. Then the two independent polarizations ϕ_{23} and $\phi_{22} - \phi_{33}$ are transformed by $S_{\alpha\beta}$ into $\Phi_{23} = -2\,\Omega\left(x^0 - x_P^0\right)(\phi_{22} - \phi_{33})/2$ and $1/2\,(\Phi_{22} - \Phi_{33}) = 2\,\Omega(x^0 - x_P^0)\phi_{23}$.

For closed integration paths and vanishing spin, Eq. (1.88) coincides with the solution of a scalar particle in a gravitational field, as expected. This proves the frequently quoted statement that gravitational radiation propagating in a gravitational background is affected by gravitation in the same way that electromagnetic radiation is (when the photon spin is neglected) [76].

The geometrical optics approximation follows immediately from (1.88) with $S_{\alpha\beta} = 0\,, T_{\alpha\beta} = 0$. From (1.86) and (1.88) one gets

$$\Phi^0_{\mu\nu} = \phi_{\mu\nu}(x) + \frac{1}{2}\int_P^x dz^\lambda \gamma_{\alpha\lambda}(z)\,\partial^\alpha\phi_{\mu\nu}(x) - \tag{1.93}$$
$$-\frac{1}{2}\int_P^x dz^\lambda\left(\gamma_{\alpha\lambda,\beta}(z) - \gamma_{\beta\lambda,\alpha}(z)\right)\left[\left(x^\alpha - z^\alpha\right)\partial^\beta\right]\phi_{\mu\nu}(x)\,.$$

The general relativistic deflection of a spin-2 particle in a gravitational field follows immediately from $\Phi^0_{\mu\nu}$. Assuming, for simplicity, that the spin-2 particles are massless and propagate along the z-direction, so that $k^\alpha \simeq (k, 0, 0, k)$, and $ds^2 = 0$

or $dt = dz$, using plane waves for $\phi_{\mu\nu}$ and writing

$$\chi = k_\sigma x^\sigma - \frac{1}{4}\int_P^x dz^\lambda(\gamma_{\alpha\lambda,\beta}(z) - \gamma_{\beta\lambda,\alpha}(z))[(x^\alpha - z^\alpha)k^\beta - (x^\beta - z^\beta)k^\alpha] + \\ + \frac{1}{2}\int_P^x dz^\lambda \gamma_{\alpha\lambda}(z)k^\alpha \,, \tag{1.94}$$

the particle momentum is

$$\tilde{k}_\sigma = \frac{\partial \chi}{\partial x^\sigma} \equiv \chi_{,\sigma} = k_\sigma - \frac{1}{2}\int_P^x dz^\lambda \left(\gamma_{\sigma\lambda,\beta} - \gamma_{\beta\lambda,\sigma}\right) k^\beta + \frac{1}{2}\gamma_{\alpha\sigma}k^\alpha \,. \tag{1.95}$$

It then follows from (1.95) that χ satisfies the eikonal equation $g^{\alpha\beta}\chi_{,\alpha}\chi_{,\beta} = 0$.

The deflection angle is particularly simple to calculate if the background metric is

$$\gamma_{00} = 2\varphi(r) \,,\; \gamma_{ij} = 2\varphi(r)\delta_{ij} \,, \tag{1.96}$$

where $\varphi(r) = -GM/r$ and $r = \sqrt{x^2 + y^2 + z^2}$, which is frequently used in gravitational lensing. For this metric, χ is given by

$$\chi \simeq k\int_P^x dz'\phi - \frac{k}{2}\mathcal{I}_k \tag{1.97}$$

$$\mathcal{I}_k \equiv \int_P^x \{(x - x')\phi_{,z'}dx' + (y - y')\phi_{,z'}dy' - 2[(x - x')\phi_{,x'} + (y - y')\phi_{,y'}]dz'\} \,.$$

The space components of the momentum are therefore

$$\tilde{k}_1 = 2k\int_P^x \left(-\frac{1}{2}\frac{\partial\varphi}{\partial z}dx + \frac{\partial\varphi}{\partial x}dz\right) \,, \tag{1.98}$$

$$\tilde{k}_2 = 2k\int_P^x \left(-\frac{1}{2}\frac{\partial\varphi}{\partial z}dy + \frac{\partial\varphi}{\partial y}dz\right) \,, \tag{1.99}$$

$$\tilde{k}_3 = k(1 + \varphi) \tag{1.100}$$

and one finds

$$\tilde{\mathbf{k}} = \tilde{\mathbf{k}}_\perp + k_3\,\mathbf{e}_3 \,, \quad \tilde{\mathbf{k}}_\perp = k_1\,\mathbf{e}_1 + k_1\,\mathbf{e}_2 \,, \tag{1.101}$$

where $\tilde{\mathbf{k}}_\perp$ is the component of the momentum orthogonal to the direction of propagation of the particles.

Since only phase differences are physical, it is convenient to choose the space–time path by placing the particle source at a distance very large relative to the dimensions of M, while the generic point is located at z along the z direction and $z \gg x, y$. Equations (1.98)–(1.100) simplify to

$$\tilde{k}_1 = 2k \int_{-\infty}^{z} \frac{\partial\varphi}{\partial x} dz = k\,\frac{2GM}{R^2}\,x\left(1+\frac{z}{r}\right), \tag{1.102}$$

$$\tilde{k}_2 = 2k \int_{-\infty}^{z} \frac{\partial\varphi}{\partial y} dz = k\,\frac{2GM}{R^2}\,y\left(1+\frac{z}{r}\right), \tag{1.103}$$

$$\tilde{k}_3 = k(1+\varphi), \tag{1.104}$$

where $R = \sqrt{x^2+y^2}$. By defining the deflection angle as

$$\tan\theta = \frac{\tilde{k}_\perp}{\tilde{k}_3}, \tag{1.105}$$

one finds

$$\tan\theta \sim \theta \sim \frac{2GM}{R}\left(1+\frac{z}{r}\right), \tag{1.106}$$

which, in the limit $z \to \infty$, yields Einstein's result

$$\theta_M \sim \frac{4GM}{R}. \tag{1.107}$$

The index of refraction can be derived from the known equation $n = \tilde{k}/\tilde{k}_0$. Choosing the direction of propagation of the particle along the x^3–axis, and using (1.95), one finds

$$n \simeq 1 + \frac{1}{k_0}\left(\chi_{,3} - \chi_{,0}\right) - \frac{m^2}{2k_0^2}\left(1 - \frac{1}{k_0}\chi_{,0}\right) \tag{1.108}$$

and, again, for $k_0 \gg m$, or for vanishing m,

$$n \simeq 1 + \frac{1}{2k_0}\mathcal{I}_n \tag{1.109}$$

$$\mathcal{I}_n \equiv -\int_P^x dz^\lambda \left(\gamma_{3\lambda,\beta} - \gamma_{\beta\lambda,3}\right)k^\beta + \gamma_{\alpha 3}k^\alpha + \int_P^x dz^\lambda \left(\gamma_{0\lambda,\beta} - \gamma_{\beta\lambda,0}\right)k^\beta - \gamma_{\alpha 0}k^\alpha .$$

In the case of the metric (1.96), one gets the known result

$$n \simeq 1 + \int_P^x dz^0 \gamma_{00,3} = 1 - \frac{2GM}{r}. \tag{1.110}$$

Problems

1.1 Derive an expression for χ by substituting Eq. (1.14) directly into Eq. (1.15) and using the Green function property $\Box_x G(x, x') = \delta^{(4)}(x - x')$.

1.2 Derive Eq. (1.61) for the spinorial connection of the metric for a rotating solid sphere of Sect. 1.3.5.

1.3 Derive an expression for the phase of a scalar particle propagating in the field of the rotating solid sphere of Sect. 1.3.5.

References

1. Hildebrandt, A.F., Saffren, M.M.: Proceedings of the 9th International Conference on Low Temperature Physics at Point A, p. 459 (1965); Hendricks, J.B., Rorschach, H.E., Jr.: ibid., p. 466; Bol, M., Fairbank, M.M.: ibid., p. 471; Zimmerman, J.E., Mercereau, J.E.: Phys. Rev. Lett. **14**, 887 (1965)
2. Colella, R., Overhauser, A.W., Werner, S.A.: Phys. Rev. Lett. **34**, 1472 (1975)
3. Werner, S.A., Staudenman, J.-L., Colella, R.: Phys. Rev. Lett. **42**, 1103 (1979)
4. Staudenman, J.-L., Werner, S.A., Colella, R., Overhauser, A.W.: Phys. Rev. A **21**, 1419 (1980)
5. Bonse, U., Wroblewski, T.: Phys. Rev. Lett. **51**, 1401 (1983)
6. Parker, L.: Phys. Rev. D **22**, 1922 (1980)
7. Tejada, J., Zysler, R.D., Molins, E., Chudnovsky, E.M.: Phys. Rev. Lett. **104**, 027202 (2010)
8. Opat, G.J.: In: Ning, H. (ed.) Proceedings of the 3rd Marcel Grossmann Meeting in General Relativity. Science Press, North-Holland, Amsterdam (1983)
9. Werner, S.A., Kaiser, H.: In: Audretsch, J., de Sabbata, V. (eds.) Quantum Mechanics in Curved Space-Time, p. 1. Plenum Press, New York (1990)
10. Riehle, F., Kister, Th., Witte, A., Helmcke, J., Bordé, Ch.J.: Phys. Rev. Lett. **67**, 177 (1991); Bordé, Ch.J.: Phys. Lett. A **140**, 10 (1989); Bordé, Ch.J., et al.: Phys. Lett. A **188**, 187 (1989); Bordé, Ch.J.: In: Berman, P. (ed.) Atom Interferometry. Academic Press, London (1997)
11. A highly instructive collection of papers on Precision Measurements in Fundamental Physics is contained. In: Blaum, K., Müller, H., Severijns, N. (eds.) Ann. der Phys. A1–A2 (2013)
12. Howl, R., Hackermüller, L., Bruschi, D.E., Fuentes, I.: Adv. Phys. **3**, 1383184 (2018)
13. MICROSCOPE Mission: Phys. Rev. Lett. **119**, 231101 (2017)
14. Hehl, F.W., Ni, W.-T.: Phys. Rev. D **42**, 2045 (1990)
15. Hehl, F.W., Lämmerzahl, C.: In: Ehlers, J., Schäfer, G. (eds.) Relativistic Gravity Research. Berlin (1992)
16. Audretsch, J., Lämmerzahl, C.: Appl. Phys. B **54**, 351 (1992)
17. Huang, J.C.: Ann. der Phys. **3**, 53 (1994)
18. Mashhoon, B.: Phys. Rev. Lett. **61**, 2639 (1988); Phys. Lett. A **139**, 103 (1989); **143**, 176 (1990); **145**, 147 (1990); Phys. Rev. Lett. **68**, 3812 (1992)
19. Singh, D., Papini, G.: Nuovo Cimento B **115**, 223 (2000)
20. Obukhov, Y.N.: Phys. Rev. Lett. **86**, 192 (2001)
21. Castro, Luis B.: Eur. Phys. J. C **76**, 62 (2016)
22. Hojman, S.A., Asenjo, F.A.: arXiv:1610.08719v1 [gr-qc], 27 Oct 2016
23. Parikh, M., Wilczek, F., Zahariade, G.: arXiv:2010.08205v1 [hep-th], 16 Oct 2020
24. Wilczek, F.: arXiv:05669v1 [con-mat.mes-hall], 19 Apr 2016

25. Ni, W.T.: In: Sato, H., Nakamura, T. (eds.) Proceedings of the 6th Marcel Grossmann Meeting on General Relativity, p. 356. World Scientific, Singapore (1992); Pan, S.-S., Ni, W.-T., Chen, S.-C.: ibid., p. 364; Jen, T.-H., Ni, W.-T., Pan, S.-S., Wang, S.-L.: ibid., p. 489; Daniels, J.M., Ni, W.-T.: ibid., p. 1629
26. Silverman, M.P.: Phys. Lett. A **152**, 133 (1991); Nuovo Cimento D **14**, 857 (1992)
27. Venema, B.J., Majumder, P.K., Lamoreaux, S.K., Hecker, B.R., Fortson, E.N.: Phys. Rev. Lett. **68**, 135 (1992)
28. Wineland, D.J., Bollinger, J.J., Heinzen, D.J., Itano, W.M., Raizen, M.G.: Phys. Rev. Lett. **67**, 1735 (1991)
29. Ritter, R.C., Winkler, L.I., Gillies, G.T.: Phys. Rev. Lett. **70**, 701 (1993)
30. Tarallo, M.G., Mazzoni, T., Poli, N., Sutyrin, D.V., Zhang, X., Tino, G.M.: arXiv:1403.1161v2 [physics.atom-ph], 23 June 2014
31. Schlippert, D., Hartwig, J., Albers, H., Richardson, L.L., Schubert, C., Roure, A., Schleich, W.P.: Phys. Rev. Lett. **112**, 203002 (2014)
32. Braginsky, V.B., Caves, C.M., Thorne, K.S.: Phys. Rev. D **15**, 2047 (1977)
33. Cai, Y.Q., Lloyd, D.G., Papini, G.: Phys. Lett. A **178**, 225 (1993)
34. Berry, M.V.: Proc. R. Soc. Lond. **A392**, 45 (1984)
35. Simon, B.: Phys. Rev. Lett. **51**, 2167 (1983)
36. Wilczek, F., Zee, A.: Phys. Rev. Lett. **52**, 2111 (1984)
37. Fock, V.: Phys. Zeits. Sowjetunion **12**, 404 (1937); Nambu, Y.: Prog. Theor. Phys. **5**, 82(1950); Feynman, R.P.: Phys. Rev. **80**, 440 (1950); Schwinger, J.: Phys. Rev. **82**, 664 (1951)
38. Cai, Y.Q., Papini, G.: Mod. Phys. Lett. A **4**, 1143 (1989)
39. Cai, Y.Q., Papini, G.: Mod. Phys. Lett. A **4**, 1143 (1989); Gen. Relativ. Gravit. **22**, 259 (1990)
40. Cai, Y.Q., Papini, G.: Class. Quantum Gravity **7**, 269 (1990)
41. Garrison, J.C., Chiao, R.Y.: Phys. Rev. Lett. **60**, 165 (1988)
42. Anandan, J.: Phys. Rev. Lett. **60**, 2555 (1988)
43. Cheng, H., Coon, D.D., Zhu, X.Q.: Phys. Rev. D **26**, 896 (1982)
44. Feng, L.L., Lee, W.L.: Int. J. Mod. Phys. D **10**, 961 (2001)
45. Ji, P., Bai, Y., Wang, L.: Phys. Rev. D **75**, 024010 (2007)
46. DeWitt, B.S.: Phys. Rev. Lett. **16**, 1092 (1966)
47. Papini, G.: Nuovo Cimento **45**, 66 (1966); Phys. Lett. **23**, 418 (1966); Nuovo Cimento B **52**, 136 (1967); Phys. Lett. A **24**, 32 (1967); Nuovo Cimento B **63**, 549 (1969)
48. Anandan, J.: Phys. Rev. Lett. **47**, 463 (1981); Chiao, R.Y.: Phys. Rev. **25**, 1655 (1982); Anandan, J.: Phys. Lett. A**110**, 446 (1985)
49. Kowitt, M.: Phys. Rev. B **49**, 704 (1994); Harris, E.G.: Found. Phys. Lett. **12**, 201 (1999)
50. Quach, J.Q.: Phys. Rev. Lett. **114**, 081104(2015); **118**, 139901 (2017); Eur. Phys. J. C **80**, 986 (2020)
51. Ummarino, G.A., Gallerati, A.: arXiv:2009.04967 v1[gr-qc], 10 Sept 2020
52. Abele, H., Jenke, T., Leeb, H., Schmiedmayer, J.: Phys. Rev. D **81**, 065019 (2010)
53. Nesvizhevsky, V.V., Börner, H.G., Petukhov, A.K., Abele, H., BaeBler, S., Rueb, F.J., Stöferle, T., Westphal, A., Gagarski, A.M., Petrov, G.A., Srelkov, A.V.: Nature **415**, 297 (2002)
54. Jenke, T., Geltenbort, P., Lemmel, H., Abele, H.: Nat. Phys. **7**, 468 (2011)
55. Cai, Y.Q., Papini, G.: Class. Quantum Gravity **6**, 407 (1989)
56. Cai, Y.Q., Papini, G.: Phys. Rev. Lett. **66**, 1259 (1991); **68**, 3811 (1992)
57. Thompson, R., Papini, G.: In: Mann, R.B., McLenaghan, R.G. (eds.) Proceedings of the 5th Canadian Conference on General Relativity and Relativistic Astrophysics, p. 459. World Scientific, Singapore (1993)
58. Corichi, A., Pierri, M.: Phys. Rev. D **51**, 5870 (1995)
59. Mostafazadeh, A.: J. Phys. A **31**, 7829 (1998)
60. Skrotskii, G.V.: Dokl. Akad. Nauk SSSR **114**, 73 (1957) [Sov. Phys. Dokl. **2**, 226 (1957)]
61. Weinberg, S.W.: Gravitation and Cosmology. Wiley, New York (1972)
62. Lense, J., Thirring, H.: Z. Phys. **19**, 156 (1918); (English translation: Mashhoon, B., Hehl, F.W., Theiss, D.S.: Gen. Relativ. Gravit. **16**, 711 (1984))

63. Lambiase, G.: Braz. J. Phys. **35**, 462–469 (2005); Lambiase, G.: Mon. Not. R. Astron. Soc. **362**, 867–871 (2005)
64. De Oliveira, C.G., Tiomno, J.: Nuovo Cimento **24**, 672 (1962)
65. Peres, A.: Suppl. Nuovo Cimento **24**, 389 (1962)
66. Weyl, H.: Z. Phys. **56**, 330 (1929)
67. Corum, J.F.: J. Math. Phys. **18**, 770 (1977)
68. Irvine, W.M.: Physica **30**, 1160 (1964)
69. Itzykson, C., Zuber, J.B.: Quantum Field Theory. McGraw-Hill Inc., New York (1980)
70. Sagnac, M.G.: C. R. Acad. Sci. (Paris) **157**, 708 (1913); **157**, 1410 (1913); Post, E.J.: Rev. Mod. Phys. **39**, 475 (1967)
71. Cnockaert, S.: Higher Spin Gauge Field Theories. arXiv:hep-th/0606121
72. Misner, C.S., Thorne, K.S., Wheeler, J.A.: Gravitation. Freeman, San Francisco (1973)
73. Takahashi, R., Nakamura, T.: Astrophys. J. **595**, 1039 (2003)
74. Shen, J.Q.: Phys. Rev. D **70**, 067501 (2004)
75. Ramos, J., Mashhoon, B.: Phys. Rev. D **73**, 084003 (2006)
76. Thorne, K.S.: In: Hawking, S.W., Israel, W. (eds.) Three Hundred Years of Gravitation, p. 361. Cambridge University Press, Cambridge (1987)

The Mashhoon Effect. Spin-Gravity Interactions

2

Abstract

The Mashhoon effect is the main subject of this chapter. Its presence and importance in muon $g - 2$ experiments are discussed together with the limits the effect places on P and T invariance. The effect is generalized to include rotating compound spin systems. The helicity and chirality precessions of fermions in external gravitational fields are also considered.

2.1 Introduction

It was Mashhoon who pointed out that the usual description of physical phenomena given by an accelerated observer, and based on a hypothesis of locality, was unduly restrictive when dealing with phenomena outside the realm of geometrical optics [1,2]. The hypothesis of locality stipulates, in fact, that an accelerated observer is locally equivalent to an instantaneously comoving inertial observer. This restricts the measurements carried out by an accelerated observer to measurements over point-like intervals in space–time and excludes phenomena involving wavelength and period of a wave, quantities that are observed over extended space–time intervals. A minimal extension of the hypothesis of locality to wave phenomena implies that terms linear in a particle wavelength represent first-order corrections to the eikonal approximation. Rotational inertia, thus Mashhoon argued, is not complete without the introduction of spin. Typically, spin-rotation coupling corresponds to a wave effect generated by the intrinsic spin of a particle. Because of spin-rotation coupling the energy of a spinning particle thus depends not only on its velocity relative to the observer, but also on the rotational acceleration of the observer.

Mashhoon's research opened up an area of quantum related gravity in which laboratory experiments can be carried out and compared with theory [3]. Mashhoon's views were further strengthened by Hehl and Ni who derived the spin-rotation interaction directly from the Dirac equation [4]. It also follows from all wave equations considered that a spin-gravity coupling exists. It is present for spin-1 particles, for

G. Lambiase and G. Papini, *The Interaction of Spin with Gravity in Particle Physics*, Lecture Notes in Physics 993, https://doi.org/10.1007/978-3-030-84771-5_2

spin-1/2 particles and again for spin-2 particles. Spin-rotation coupling has been observed for photons and neutrons [5–7]. Spin-gravity coupling plays, moreover, a wider role in physics [8], starting with the classical electron [9]. It is present, as shown below, in g-2 experiments and helicity transitions in neutrino physics and can be used to set stringent limits on the invariance of discreet symmetries.

The direct action of the Mashhoon coupling term on spin motion and its importance in spintronics are discussed in Chap. 6. Mashhoon's studies of the interaction of the inertia of intrinsic spin have fostered attempts to elucidate the similarities between gravitation and electromagnetism, a research area also known as gravitoelectromagnetism [10].

Mashhoon's concerns about the use of the locality hypothesis also extend to measuring processes [1,11]. For standard, accelerated measuring devices, for instance, substituting inertial frames for non-inertial ones entail the introduction of an upper limit to the proper acceleration of a particle, or maximal acceleration, itself the subject of much research [12,13].

2.2 Spin-Rotation Coupling

Using the results of Sects. (1.3.4) and (1.3.5), it is now possible to derive the Hamiltonian of a fermion subjected to rotation and acceleration. It is obtained by isolating the time derivative in the Dirac equation. The result is

$$H = \boldsymbol{\alpha}\cdot\mathbf{p} + m\beta + V(x)$$
$$V(x) = \frac{1}{2}[(\mathbf{a}\cdot\mathbf{x})(\mathbf{p}\cdot\boldsymbol{\alpha}) + (\mathbf{p}\cdot\boldsymbol{\alpha})(\mathbf{a}\cdot\mathbf{x})] + m\beta(\mathbf{a}\cdot\mathbf{x}) - \boldsymbol{\omega}\cdot\left(\mathbf{L}+\frac{\sigma}{2}\right), \quad (2.1)$$

where $\mathbf{L}$ is the orbital angular momentum and $\boldsymbol{\sigma}$ represents the usual Pauli matrices. The first three terms in $V(x)$ represent relativistic energy–momentum effects. The term $-\boldsymbol{\omega}\cdot\mathbf{L}$ is a Sagnac-type effect. The last term, $-\frac{1}{2}\boldsymbol{\omega}\cdot\boldsymbol{\sigma}$, is the spin-rotation coupling, or Mashhoon effect. The non-relativistic effects can be obtained by applying three successive Foldy–Wouthysen transformations to H [14]. One obtains to lowest order (Appendix B)

$$H = m\beta + \beta\frac{p^2}{2m} + \beta m(\mathbf{a}\cdot\mathbf{x}) + \frac{\beta}{2m}\mathbf{p}(\mathbf{a}\cdot\mathbf{x})\cdot\mathbf{p} - \boldsymbol{\omega}\cdot\left(\mathbf{L}+\frac{\sigma}{2}\right) + \frac{1}{4m}\sigma\cdot(\mathbf{a}\times\mathbf{p}). \quad (2.2)$$

The third term in Eq. (2.2) is the energy–momentum effect observed by Bonse and Wroblewski [15]. The term $-\boldsymbol{\omega}\cdot\mathbf{L}$ was predicted by Page [16] and observed by Werner and collaborators [17]. Hehl and Ni [4] have re-derived all terms starting from the covariant Dirac equation. The fourth term is a kinetic energy effect, and the last term represents the spin–orbit coupling [4]. The Mashhoon effect can also be derived from the gravitational Berry phases of Sects. (1.3.2), (1.3.3), (1.3.4) and (1.3.6) [18]. As discussed by Mashhoon, the effect violates the hypothesis of locality, which is valid for classical point-like particles and optical rays and is widely used in relativity. The effect also violates the equivalence principle because it does not couple universally to matter [1,3]. Direct experimental verifications of the Mashhoon effect have been reported [5–7]. It is shown in the following sections that the Mashhoon effect also plays an essential role in measurements of the anomalous magnetic moment, or $g-2$ factor of the muon and can be used to set stringent limits on the invariance of discreet symmetries. Its truly quantum mechanical and pervasive nature is also attested by the description of compound spin systems accelerating in a storage ring given in the following.

2.2.1 Spin-Rotation Coupling in Muon g-2 Experiments

Precise measurements of the $g-2$ factor involve muons in storage rings which consist of a vacuum tube, a few metres in diameter, in a uniform, vertical magnetic field **B**. Muons on equilibrium orbits within a small fraction of the maximum momentum are almost completely polarized with spin vectors pointing in the direction of motion. As the muons decay, the highest energy electrons with spin almost parallel to the momentum, are projected forward in the muon rest frame and are detected around the ring. Their angular distribution does therefore reflect the precession of the muon spin along the cyclotron orbits [19,20]. It is convenient to start from the covariant Dirac equation of Sect. (1.3.4) and to use the chiral representation for the usual Dirac matrices

$$\gamma^0=\beta=\begin{pmatrix}0 & -I\\ -I & 0\end{pmatrix},\quad \gamma^i=\begin{pmatrix}0 & \sigma^i\\ -\sigma^i & 0\end{pmatrix},\quad \alpha^i=\begin{pmatrix}\sigma^i & 0\\ 0 & -\sigma^i\end{pmatrix},$$
$$\sigma^{0i}=i\begin{pmatrix}\sigma^i & 0\\ 0 & -\sigma^i\end{pmatrix},\quad \sigma^{ij}=\epsilon_k^{ij}\begin{pmatrix}\sigma^k & 0\\ 0 & \sigma^k\end{pmatrix},\quad \gamma^5=\begin{pmatrix}I & 0\\ 0 & -I\end{pmatrix}. \qquad (2.3)$$

One must now add to the Hamiltonian the effect of a magnetic field **B** on the total (magnetic plus anomalous) magnetic moment of the particle. Assuming for simplicity that all quantities in H are time-independent and referring them to a left-handed triad of axes comoving with the particle in the x_3-direction and rotating in the x_2-direction, one finds

$$H=\alpha^3 p_3+m\beta+\frac{1}{2}[-a_1R(\alpha^3p_3)-(\alpha^3p_3)a_1R]+\beta m a_1 R-\boldsymbol{\omega}\cdot\mathbf{L}-$$
$$\frac{1}{2}\omega_2\sigma^2+\mu B\sigma^2\equiv H_0+H', \qquad (2.4)$$

where $B_2=-B$, $\mu=(1+\frac{g-2}{2})\mu_0$, $\mu_0=\frac{e\hbar}{2mc}$ is the Bohr magneton, $H'=-\frac{1}{2}\omega_2\sigma^2+\mu\beta\sigma^2$ and R is the radius of the muon's orbit. Electric fields used to stabilize the orbits and stray radial electric fields can also affect the muon spin. Their effects can be cancelled by choosing an appropriate muon momentum and will be neglected in what follows. Before decay, the muon states can be represented in the form

$$|\psi(t)>=a(t)|\psi_+>+b(t)|\psi_->, \qquad (2.5)$$

where $|\psi_+>$ and $|\psi_->$ are the right and left helicity states of H_0. Substituting (2.5) into the Schroedinger equation one obtains

$$i\frac{\partial}{\partial t}\begin{pmatrix} a \\ b \end{pmatrix} = \begin{pmatrix} <\psi_+|(H_0+H')|\psi_+> & <\psi_+|H'|\psi_-> \\ <\psi_-|H'|\psi_+> & <\psi_-|(H_0+H')|\psi_-> \end{pmatrix}\begin{pmatrix} a \\ b \end{pmatrix}$$
$$= \begin{pmatrix} E - i\frac{\Gamma}{2} & i(\frac{\omega_2}{2} - \mu B) \\ -i(\frac{\omega_2}{2} - \mu B) & E - i\frac{\Gamma}{2} \end{pmatrix} \equiv M \begin{pmatrix} a \\ b \end{pmatrix}, \quad (2.6)$$

where Γ represents the width of the muon. Notice that the spin-rotation coupling is off-diagonal in (2.6). This is an indication that the Mashhoon effect violates the equivalence principle [1]. The matrix M can be diagonalized. Its eigenvalues are $h_1 = E - i\frac{\Gamma}{2} + (\frac{\omega_2}{2}) - \mu B)$, $h_2 = E - i\frac{\Gamma}{2} - (\frac{\omega_2}{2} - \mu B)$, with the corresponding eigenvectors

$$|\psi_1>= \frac{1}{\sqrt{2}}[i|\psi_+> + |\psi_->]; \; |\psi_2>= \frac{1}{\sqrt{2}}[-i|\psi_+> + |\psi_->]. \quad (2.7)$$

The solution of Eq. (2.6) is therefore

$$|\psi(t)> = \frac{1}{\sqrt{2}}(e^{-ih_1t}|\psi_1> + e^{-ih_2t}\psi_2>) =$$
$$\frac{1}{2}[(ie^{-ih_1t} - ie^{-ih_2t})|\psi_+> + (e^{-ih_1t} + e^{-ih_2t})|\psi_->], \quad (2.8)$$

where $|\psi(0)>= |\psi_->$. The spin-flip probability is

$$P_{\psi_-\to\psi_+} = |<\psi_+|\psi>|^2 = \frac{e^{-\Gamma t}}{2}[1 - \cos(2\mu B - \omega_2)t], \quad (2.9)$$

where the Γ-term accounts for the observed exponential decrease in electron counts due to the loss of muons by radioactive decay [20]. The spin-rotation contribution to $P_{\psi\to\psi_+}$ is represented by ω_2 which is the cyclotron angular velocity $\frac{eB}{m}$. The spin-flip angular frequency is then

$$\Omega = 2\mu B - \omega_2 = (1 + \frac{g-2}{2})\frac{eB}{m} - \frac{eB}{m} = \frac{g-2}{2}\frac{eB}{m}, \quad (2.10)$$

which is precisely the observed modulation frequency of the electron counts [21]. This result is independent of the value of the anomalous magnetic moment of the particle. It is therefore the Mashhoon effect that gives prominence to the $g-2$ term in Ω by exactly cancelling, in $2\mu B$, the much larger contribution μ_0 that comes from fermions when the anomalous magnetic moment is neglected [22].

2.2.2 Spin-Rotation Coupling and Limits on P and T Invariance

A discrepancy $a_\mu(\exp) - a_\mu(SM) = 43 \times 10^{-10}$ has been observed [23,24] between the experimental and standard model values of the muon's anomalous g value, $a_\mu = \frac{g-2}{2}$. This discrepancy can be used to set an upper limit on P and T invariance violations in spin-rotation coupling.

The possibility that discrete symmetries in gravitation be not conserved has been considered by some authors [25–28]. Attention has in general been paid to the potential

$$U(r) = \frac{GM}{r}\left[\alpha_1 \boldsymbol{\sigma} \cdot \hat{r} + \alpha_2 \boldsymbol{\sigma} \cdot \boldsymbol{v} + \alpha_3 \hat{r} \cdot (\boldsymbol{v} \times \boldsymbol{\sigma})\right], \tag{2.11}$$

which applies to a particle of generic spin $\boldsymbol{\sigma}$. The first term, introduced by Leitner and Okubo [26], violates the conservation of P and T. The same authors determined the upper limit $\alpha_1 \leq 10^{-11}$ from the hyperfine splitting of the ground state of hydrogen. The upper limit $\alpha_2 \leq 10^{-3}$ was determined in [28] from SN 1987A data. The corresponding potential violates the conservation of P and C. Conservation of C and T is violated by the last term, while (2.11), as a whole, conserves CPT. There is, as yet, no upper limit on α_3. These studies are extended here to the Mashhoon term.

Assume that all quantities in the effective Hamiltonian are time-independent and referred to a left-handed triad of axes comoving with the muons and rotating about the x_2-axis in the clockwise direction of motion of the muons. The x_3-axis is tangent to the orbits and in the direction of the muon momentum. The magnetic field is $B_2 = -B$. Of all the terms that appear in the Dirac Hamiltonian, only the Mashhoon term couples the helicity states of the muon. The remaining terms contribute to the overall energy E of the states and the corresponding part of the Hamiltonian is indicated by H_0 [29].

If it is now assumed that the coupling of rotation to $|\,\psi_+ >$ differs in strength from that to $|\,\psi_- >$, then the Mashhoon term can be altered by means of a matrix

$$A = \begin{pmatrix} \kappa_1 & 0 \\ 0 & \kappa_2 \end{pmatrix}$$

that reflects the hypothetically different coupling of rotation to the two helicity states. The total effective Hamiltonian is $H_{eff} = H_0 + H'$, where

$$H' = -\frac{1}{2}A\omega_2\sigma_2 + \mu B\sigma_2, \tag{2.12}$$

$\mu = (1 + a_\mu)\mu_0$ represents the total magnetic moment of the muon and μ_0 is the Bohr magneton. A violation of P and T in (2.12) would arise through $\kappa_2 - \kappa_1 \neq 0$. The constants κ_1 and κ_2 can differ from unity by small amounts ϵ_1 and ϵ_2.

The coefficients $a(t)$ and $b(t)$ in (2.5) still evolve in time according to

$$i\frac{\partial}{\partial t}\begin{pmatrix} a(t) \\ b(t) \end{pmatrix} = M\begin{pmatrix} a(t) \\ b(t) \end{pmatrix}, \tag{2.13}$$

but now M is given by

$$M = \begin{pmatrix} E - i\frac{\Gamma}{2} & i\left(\kappa_1\frac{\omega_2}{2} - \mu B\right) \\ -i\left(\kappa_2\frac{\omega_2}{2} - \mu B\right) & E - i\frac{\Gamma}{2} \end{pmatrix}, \tag{2.14}$$

where Γ again represents the width of the muon. The spin-rotation term is off-diagonal in (2.14) and does not therefore couple to matter universally. It violates Hermiticity [30]. It also violates T, P and PT, as stated, while nothing can be said about CPT conservation which requires M to be Hermitean [31,32]. Because of the non-Hermitean nature of (2.12), one expects Γ itself to be non-Hermitean. The resulting corrections to the width of the muon are, however, of second order in the ϵ's and are neglected.

M now has eigenvalues

$$\begin{aligned} h_1 &= E - i\frac{\Gamma}{2} + R \\ h_2 &= E - i\frac{\Gamma}{2} - R, \end{aligned} \tag{2.15}$$

where

$$R = \sqrt{\left(\kappa_1\frac{\omega_2}{2} - \mu B\right)\left(\kappa_2\frac{\omega_2}{2} - \mu B\right)}, \tag{2.16}$$

and eigenstates

$$\begin{aligned} |\psi_1 > &= b_1\left[\eta_1|\psi_+ > + |\psi_- >\right], \\ |\psi_2 > &= b_2\left[\eta_2|\psi_+ > + |\psi_- >\right]. \end{aligned} \tag{2.17}$$

One also finds

$$\begin{aligned} |b_1|^2 &= \frac{1}{1 + |\eta_1|^2} \\ |b_2|^2 &= \frac{1}{1 + |\eta_2|^2} \end{aligned} \tag{2.18}$$

and

$$\eta_1 = -\eta_2 = \frac{i}{R}\left(\kappa_1\frac{\omega_2}{2} - \mu B\right). \tag{2.19}$$

Then the muon states (2.9) are

$$|\psi(t) >= \frac{1}{2}e^{-iEt - \frac{\Gamma t}{2}}\left[-2i\eta_1 \sin Rt|\psi_+ > +2\cos Rt|\psi_- >\right], \tag{2.20}$$

where the condition $|\psi(0)>=|\psi_->$ has been applied. The spin-flip probability is therefore

$$\begin{aligned} P_{\psi_-\to\psi_+} &= |<\psi_+|\psi(t)>|^2 \\ &= \frac{e^{-\Gamma t}}{2}\frac{\kappa_1\omega_2 - 2\mu B}{\kappa_2\omega_2 - 2\mu B}[1-\cos 2Rt]. \end{aligned} \quad (2.21)$$

When $\kappa_1 = \kappa_2 = 1$, Eq. (1.59) yields [29]

$$P_{\psi_-\to\psi_+} = \frac{e^{-\Gamma t}}{2}\left[1-\cos\left(a_\mu\frac{eB}{m}t\right)\right], \quad (2.22)$$

that provides the appropriate description of the spin-rotation contribution to the spin-flip transition probability. Notice that the case $\kappa_1 = \kappa_2 = 0$ (no spin-rotation coupling) yields

$$P_{\psi_-\to\psi_+} = \frac{e^{-\Gamma t}}{2}\left[1-\cos(1+a_\mu)\frac{eB}{m}\right] \quad (2.23)$$

and does not therefore agree with the results of the $g-2$ experiments. Hence the necessity of accounting for spin-rotation coupling whose contribution cancels the factor $\frac{eB}{m}$ in (2.10) [29].

Substituting $\kappa_1 = 1+\epsilon_1, \kappa_2 = 1+\epsilon_2$ into (2.21), one finds

$$P_{\psi_-\to\psi_+} \simeq \frac{e^{-\Gamma t}}{2}\left[1-\cos\frac{eB}{m}(a_\mu-\epsilon)t\right], \quad (2.24)$$

where $\epsilon = \frac{1}{2}(\epsilon_1+\epsilon_2)$. One may attribute the discrepancy between the experimental value $a_\mu(exp)$ and the standard model value $a_\mu(SM)$ to a violation of the conservation of the discrete symmetries by the spin-rotation coupling term. The upper limit on the violation of P, T and PT is derived from (2.24) assuming that the deviation from the current value of $a_\mu(SM)$ is wholly due to ϵ. The upper limit is therefore 43×10^{-10}.

Some more information can be extracted from a_μ data. One may in fact assume that the coupling of rotation to the two helicity states of the fermion is opposite. In this case the parameters have values $\kappa_1 = 1, \kappa_2 = -1$. This is the anti-Hermitean limit of the interaction. The oscillation frequency is then

$$R = \frac{1}{2}\sqrt{(2\mu B)^2 - \omega_2^2} = \frac{eB}{2m}\sqrt{2a_\mu + a_\mu^2} \quad (2.25)$$

and Eq. (2.21) gives

$$P_{\psi_-\to\psi_+} \simeq \frac{e^{-\Gamma t}}{2}\frac{a_\mu}{2+a_\mu}\left[1-\cos\left(\frac{eB}{m}\sqrt{2a_\mu}t\right)\right]. \quad (2.26)$$

Equations (2.26) and (2.22) differ in amplitude and frequency. In fact the amplitude of (2.26) is much smaller than that of (2.22) while its frequency is higher than that actually observed. The choice $\kappa_1 = 1, \kappa_2 = -1$ is not therefore supported experimentally.

It also follows from (2.12), (2.13) and (2.14) that the weight of a rotating object depends on its direction of rotation. The problem has been studied experimentally in [33]. An upper limit on this effect can be obtained in the present model from (2.12). The eigenstate energy difference due to spin-rotation coupling is in fact

$$-i\frac{\omega_2}{2}(<\psi_-|\sigma^2|\psi_+> + <\psi_+|\sigma^2|\psi_->) = \frac{\omega_2}{2}(1+\epsilon). \tag{2.27}$$

The additional energy difference is therefore $\frac{\epsilon\omega_2}{2}$, where $\epsilon = -43 \times 10^{-10}$. Returning to normal units, the corresponding decrease in mass for a muon of positive helicity is $\Delta m = -\frac{\epsilon e B \hbar}{mc^2} \simeq -3.1 \times 10^{-48} g$ and the decrease in weight is $g\Delta m \simeq 4 \times 10^{-45} dyne$.

The fraction of total rotational energy associated with the effect is $\epsilon/2$. If one applies this result to all the particles of the gyroscope used in the experiment of [33], then one finds that the energy difference of the two rotation states of the body is at most $\frac{\epsilon}{2}\frac{1}{2}I\omega^2 \simeq 2.4\, erg$, corresponding to a change in mass $\leq 2.6 \times 10^{-21} g$ and a change in weight $\leq 2.6 \times 10^{-18} dyne$, in agreement with the null experimental results of [34,35].

More generally, the recent observation of high-energy Compton spectra [36] seems to rule out the possibility that anti-matter behave anti-gravitationally with a confidence level close to 100%. The same data at 13 GeV suggest a C violation by a maximal amount of $(9 \pm 2) \times 10^{-12}$ and a P violation by $(13 \pm 2) \times 10^{-12}$. The data asymmetry implies a stronger gravitational coupling to left helicity electrons than to right helicity positrons.

New Results from $g - 2$ Collaboration

Recent results by the Muon $g - 2$ collaboration [37] have improved the accuracy of the measurements of the μ^+ anomalous magnetic moment to the value $0.46 ppm$, which, combined with previous measurements of a_μ of both μ^+ and μ^-, yield the new experimental average

$$a_\mu = 116592061(41) \times 10^{-11} (0.37\,\text{ppm})$$

and increase the discrepancy between experiment and theory to 4.2 standard deviations (Fig. 2.1). While there still is a contribution of 7116 ± 184, due to lattice hadronic vacuum polarization, which has not been used in the present calculations [38] to account for the discrepancy, it is interesting to assume, as in Sect. 2.2.2, that the discrepancy be due to a difference in coupling between rotation and spin of

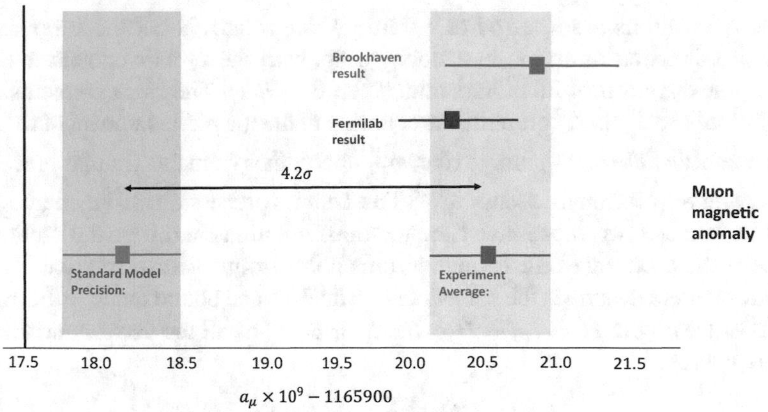

Fig. 2.1 The Muon $g-2$ experiment at Fermilab has confirmed the result obtained from the experiment performed at Brookhaven National Lab. These results show the discrepancy of muon physics from the Standard Model prediction

μ^+, μ^-. This leads to an upper limit in the violation of conservation of P, T and $PT \leq 41 \times 10^{-11}$(0.35 ppm), still larger than the limit $\alpha_1 \leq 10^{-11}$ found by Leitner and Okubo.

2.3 Spin-Rotation Coupling in Compound Spin Objects

Studies of the effect of spin-rotation coupling and Thomas precession can be extended to bound particles accelerating in a storage ring in which there is a magnetic field perpendicular to the circular orbits of the particles. The case studied refers to a muon bound to a nucleus and dragged by it along the nucleus path in a storage ring.

It is useful to examine once more the role of spin-rotation coupling in the determination of anomalous magnetic moments. A relevant point of the $g-2$ experiment is that muons on equilibrium orbits within a small fraction of the maximum momentum are almost completely polarized with spin vectors pointing in the direction of motion. As the muons decay, those electrons projected forward in the muon rest frame can therefore be detected around the ring and their angular distribution reflects the precession of the muon spin along the cyclotron orbits.

Recently, the possibility of studying the evolution of heavily charged ions in storage rings [39,40] raises the question of the role of spin-rotation coupling in rotating compound spin systems that have applications in fields like nuclear physics, QED, bound state QED and stellar nucleosynthesis [41].

The purpose of this section is to investigate spin-rotation coupling in the semiclassical system represented by a nucleus plus a decaying charged particle, specifically a negative muon, rotating in a storage ring. The problem is closely related to the original $g-2$ experimental setup [19,20]. Though the muon is not just a more massive electron, information about the particle that is dragged along the orbit by the nucleus can be obtained only if the particle itself decays.

The calculations are referred to a frame which rotates about the x_3-axis in the clockwise direction of an ion in a storage ring, with the x_2-axis tangent to the ion orbit in the direction of its momentum. Then $\mathbf{B} = B\hat{\mathbf{u}}_3$. The main elements of the calculation are ω_{g_μ} that represents the coupling of the magnetic moment of the muon to the magnetic field in the storage ring, $\omega_{Th}^{(\mu)}$ that comes from the Thomas precession, and the muon cyclotron frequency $\omega_c^{(\mu)}$. The Thomas precession is a kinematic effect. It occurs because two successive Lorentz transformations in different directions are equivalent to a Lorentz transformation plus a three-dimensional rotation. The full Hamiltonian that describes the behaviour of a nucleus and bound muon in the external field $\mathbf{B}$ of the ring is $H = H_0 + H_1$, where H_0 contains all the usual standard terms and H_1 is [42]

$$H_1 = -\mathcal{A}\mathbf{s}\cdot\mathbf{I} - \mathbf{s}\cdot\Omega_\mu - \mathbf{I}\cdot\Omega_n\,, \tag{2.28}$$

where $\mathcal{A} \simeq Z^3 4\alpha^4 g_n/3 \sim N \times 10^{14} Hz$, $N \sim O(1)$ is the strength of the spin–spin coupling. The quantity of interest is the angular precession frequency Ω_μ given by

$$\Omega_\mu \equiv \omega_{Th}^{(\mu)} + \omega_{g_\mu} - \omega_c^{(\mu)}\,. \tag{2.29}$$

This quantity can be calculated and is

$$\omega_{Th}^{(\mu)} = \frac{e\mathbf{B}}{m}\frac{1}{\gamma_{\mu|n}\gamma_n} I_\mu - \frac{Q\mathbf{B}}{M}\frac{1}{\gamma_n} I_Q\,, \tag{2.30}$$

where the relativistic factors β_n, γ_n refer to the motion of the nucleus in the storage ring and $\beta_{\mu|n}, \gamma_{\mu|n}$ to that of the muon relative to the nucleus. The other definitions are

$$I_\mu \equiv \frac{(\gamma_{\mu|n}\gamma_n)^2\left(\beta_n^2 + \frac{\beta_{\mu|n}^2}{\gamma_n} + 2\Pi + \frac{\gamma_n\Pi^2}{\gamma_n+1} - Y\right)}{(1+\Pi)[\gamma_{\mu|n}\gamma_n(1+\Pi)+1]}\,, \tag{2.31}$$

$$I_Q \equiv \frac{(\gamma_{\mu|n}\gamma_n)^2\left[\beta_n^2\left(1 + \frac{\gamma_n\Pi}{\gamma_n+1}\right)^2 - X\right]}{\gamma_{\mu|n}\gamma_n(1+\Pi)+1}\,,$$

$$Y \equiv \frac{\beta_{\mu|n}^2[\gamma_n(2-\cos^2\theta) - \sin^2\theta]}{3\gamma_n^2}\,, \qquad X \equiv \frac{\beta_{\mu|n}^2\beta_n^2\sin^2\theta}{3(\gamma_n+1)}\,,$$

where $\Pi = \boldsymbol{\beta}_n \cdot \boldsymbol{\beta}_{\mu|n} = \beta_n\beta_{\mu|n}\cos\theta$. Equation (2.30) for ω_{Th} reduces to the standard angular frequency of the muon moving in the storage ring when $Y = 0$ (because $(\mathbf{B}\cdot\boldsymbol{\beta}_{\mu|n})\beta_{\mu|n} = 0$). In fact, in the absence of the charged nucleus ($Q = 0$), which implies $\beta_{\mu|n} = 0$, $\gamma_{\mu|n} = 1$, hence $\beta = \beta_n$, $\gamma_n = \gamma$, one re-obtains from (2.30) the standard result [43]

$$\omega_{Th,l}^{(\mu)} = \frac{eB}{m}\frac{\gamma-1}{\gamma}\,. \tag{2.32}$$

The same result is also recovered by setting $\beta_n = 0$, so that $\beta = \beta_{\mu|n}$ and $\gamma = \gamma_{\mu|n}$.

Muons have an intrinsic magnetic momentum given by $\boldsymbol{\mu}_\mu = -g_\mu \mu_B \mathbf{s}$, where g_μ is the g-factor, μ_B the Bohr magneton and $\mathbf{s}$ the muon spin. When placed in an external magnetic field $\mathbf{B}$, muons acquire an additional potential energy given by $-\boldsymbol{\mu}_\mu \cdot \mathbf{B}$. Following [42], one finds

$$\boldsymbol{\omega}_{g_\mu} = -\frac{g_\mu e}{2m}\Upsilon \mathbf{B}\,, \tag{2.33}$$

where

$$\Upsilon \equiv 1 - \frac{\gamma_{\mu|n}^2 (\boldsymbol{\beta}_{\mu|n} \cdot \hat{u}_3)^2}{\gamma(\gamma+1)}\,. \tag{2.34}$$

The calculation of g-factors, based on bound state QED, can be carried out with accuracy even when the expansion parameter is $Z\alpha \simeq 0.4$. The results agree with available direct measurements [44]. In particular, the bound state QED calculation given in [45] includes radiative corrections of order α/π and exact binding corrections. It yields

$$g_\mu = 2\left[\frac{1 + 2\sqrt{1-(\alpha Z)^2}}{3} + \frac{\alpha}{\pi} C^{(2)}(\alpha Z)\right]\,, \tag{2.35}$$

where

$$C^{(2)}(\alpha Z) \simeq \frac{1}{2} + \frac{1}{12}(\alpha Z)^2 + \frac{7}{2}(\alpha Z)^4\,.$$

In the limit $Q = 0$ one gets

$$\boldsymbol{\omega}_{g_\mu,l} = -\frac{e\boldsymbol{B} g_\mu}{2m}\,. \tag{2.36}$$

The cyclotron frequency of the bound muon is

$$\omega_c^{(\mu)} = \left[-\frac{eB}{m}\frac{\beta_n}{\beta}\frac{1-(\beta_{\mu|n}\cdot\hat{u}_1)^2}{\gamma_{\mu|n}\gamma_n(1+\Pi)^2} + \frac{QB}{M}\frac{\beta_n}{\beta}\frac{1}{\gamma_n(1+\Pi)}\left(1+\frac{\gamma_n\Pi}{\gamma_n+1}\right)\right]\hat{u}_3 \tag{2.37}$$

which, in the usual limit $Q = 0$, yields

$$\boldsymbol{\omega}_{c,l}^{(\mu)} = \frac{e\mathbf{B}}{m\gamma}\,, \tag{2.38}$$

and from (2.32), (2.36) and (2.38) one gets, in the same limit,

$$\boldsymbol{\Omega}_{\mu,l} = \boldsymbol{\omega}_{Th,l}^{(\mu)} + \boldsymbol{\omega}_{g_\mu,l} - \boldsymbol{\omega}_{c,l}^{(\mu)} = -\frac{e\mathbf{B}}{m}\left(-\frac{\gamma-1}{\gamma} + \frac{g_\mu}{2} - \frac{1}{\gamma}\right)\,. \tag{2.39}$$

The analysis of (2.38) reveals that $\boldsymbol{\omega}_{c,l}^{(\mu)}$ corresponds to the spin-rotation coupling. It therefore follows that $\boldsymbol{\omega}_c^{(\mu)}$ is the desired *generalization of the spin-rotation coupling to rotating compound spin systems.*

The angular frequencies $\omega_{Th}^{(\mu)}$, ω_{g_μ} and $\omega_c^{(\mu)}$ yield the final expression of the angular precession frequency Ω_μ

$$\boldsymbol{\Omega}_\mu \equiv \boldsymbol{\omega}_{g_\mu} + \boldsymbol{\omega}_{Th}^{(\mu)} - \boldsymbol{\omega}_c^{(\mu)} = \tag{2.40}$$

$$= -\frac{e\mathbf{B}}{m_\mu}\left(\frac{g_\mu}{2}\Upsilon - \frac{I_\mu}{\gamma_{\mu|n}\gamma_n} - U\right) - \frac{Q\mathbf{B}}{M}\frac{I_Q + V}{\gamma_n},$$

where Υ is defined in (2.34) and

$$U \equiv \frac{1-(\beta_{\mu|n}\cdot\hat{u}_1)^2}{\gamma_{\mu|n}\gamma_n(1+\Pi)^2}\frac{\beta_n}{\beta}, \tag{2.41}$$

$$V = \frac{\beta_n}{\tilde{\beta}}\frac{1}{(1+\Pi)}\left(1+\frac{\gamma_n\Pi}{\gamma_n+1}\right).$$

Since in this simplified case the magnetic field components along the x_2 and x_3-axes are neglected, $\mathbf{B} = B\hat{u}_3$, Eq. (2.40) can be re-cast in the form

$$\boldsymbol{\Omega}_\mu = \Omega_\mu\hat{u}_3, \tag{2.42}$$

where

$$\Omega_\mu \equiv -\frac{eB}{m}\left[\left(\frac{g_\mu}{2}\Upsilon\epsilon_1 - \epsilon_2\frac{I_\mu}{\gamma_{\mu|n}\gamma_n} - U\epsilon_3\right) + \frac{Z}{A}\frac{m}{m_p}\frac{\epsilon_4 I_Q + \epsilon_5 V}{\gamma_n}\right], \tag{2.43}$$

$m_p \simeq 0.9$GeV is the proton mass and the ϵ's are tags introduced to distinguish the various contributions. In particular ϵ_1 tags the g_μ contribution, ϵ_2 and ϵ_4 those due to $\omega^\mu_{(Th)}$, while ϵ_3 and ϵ_5 refer to $\omega_c^{(\mu)}$. On carrying out averages over the angles and taking $\gamma_n \sim 1.6$, one obtains

$$\Omega_\mu = \frac{1}{3T}1068B\left\{0.65548\,\frac{Z}{A}\left(\epsilon_5\frac{0.755213}{\sqrt{1-\frac{0.429654}{\gamma_{\mu|n}^2}}} + \epsilon_4\frac{0.1752+1.15226\gamma_{\mu|n}^2}{1+1.5256\gamma_{\mu|n}}\right)\right.$$

$$+\,8.86788\left[-\frac{\epsilon_3\left(0.165009+0.330018\gamma_{\mu|n}^2\right)}{\sqrt{1-\frac{0.429654}{\gamma_{\mu|n}^2}}\,\gamma_{\mu|n}^3} - \right.$$

$$-\frac{0.65548\,\epsilon_2\left(-0.841867+2.16932\gamma_{\mu|n}^2\right)}{\gamma_{\mu|n}\left(1.+1.5256\gamma_{\mu|n}\right)} +$$

Table 2.1 Partial and total contributions to Ω_μ and comparison of $^4He^{1+}$ with a few ions for which $Z/A \sim 1/2$

ϵ(Hz)	$^4He^{1+}$	$^{142}Pr^{60+}$	$^{140}Pm^{58+}$	$^{122}I^{52+}$
ϵ_1	5.6374 10^7	5.1851 10^7	5.1534 10^7	5.2735 10^7
ϵ_2	−1.9405 10^7	−1.9929 10^7	−2.0122 10^7	−1.9383 10^7
ϵ_3	−3.6903 10^7	−3.1395 10^7	−3.0947 10^7	−3.2708 10^7
ϵ_4	1.0940 10^6	0.8968 10^6	0.9220 10^6	0.9066 10^6
ϵ_5	2.0811 10^6	1.7345 10^6	1.7605 10^6	1.8105 10^6
Ω_μ(Hz)	3.2417 10^6	3.1591 10^6	3.1475 10^6	3.3613 10^6

$$+\,\epsilon_1\left(1+a_\mu\right)\left(1-\frac{0.218493\left(-1+\gamma_{\mu|n}^2\right)}{\gamma_{\mu|n}\left(1+1.5256\gamma_{\mu|n}\right)}\right)\Bigg]\Bigg\}\,. \tag{2.44}$$

The $\epsilon_1, \epsilon_2, \epsilon_3$ terms tend to balance one another (Table 2.1). The larger contributions come from ϵ_4 and ϵ_5 and remain dominant so far as $Z/A \sim 0.5$. The last two terms are not present in the original derivation of the spin-rotation coupling in which the fermion is not bound. They are entirely due to the presence of the nucleus to which the muon is attached and their contributions, as mentioned above, can be traced back to ω_{Th}^{μ} and $\omega_C^{(\mu)}$.

Equivalent generalizations of the spin-rotation coupling can also be obtained for bound bosons by means of the procedure outlined above.

Spin-rotation coupling has thus been extended to rotating, compound spin systems. The generalization has been obtained by means of successive Lorentz transformations on the usual assumption that there exists an infinity of locally inertial observers and that, therefore, the time scale over which the process takes place is small relative to the acceleration time scale of the observer, as pointed out in [1].

The Lorentz factor $\gamma_{\mu|n}$ that appears in (2.43) and (2.44) is the only free parameter of the entire calculation. It cannot, in fact be determined by comparing (2.44) with experimental data that, as yet, do not exist. One can however make some estimates. Choosing $\gamma_{\mu|n}$ equal to the Bohr atom value, given by

$$\gamma=\sqrt{1+\frac{(Z\alpha)^4}{4}+\frac{(Z\alpha)^2}{2}}\,, \tag{2.45}$$

we obtain the value $\gamma_{\mu|n} = 1.00011$ that substituted in (2.44) gives the results reported in the table.

The same value of $\gamma_{\mu|n}$ leads to the Bohr atom energy

$$E = -m(1-\gamma_{\mu|n}+(\alpha Z)^2) = -10.8952\,KeV$$

for He in good agreement with the relativistic value

$$E_R = -\frac{m(Z\alpha)^2}{2n^2}\left[1+\frac{(Z\alpha)^2}{n}\left(1-\frac{3}{4n}\right)\right] = -11.2755KeV\,.$$

It appears from the table that a typical modulation frequency $\sim 3.5 \times 10^6$Hz is superimposed on the usual exponential decay of the muon dragged along the ion orbit, while the muon polarization is approximately $\beta \sim \beta_n \sim 0.75$. As a comparison, the typical modulation in the $g-2$ experiment is $\Omega_\mu^{(g-2)} \sim 2.3 \times 10^5$Hz with a polarization $\beta^{(g-2)} \sim 0.9$.

The calculation given suggests that spin-rotation coupling can generate oscillations in a decay process and underlines the importance that ion accelerators may assume for fundamental physics. It links quantum phenomena, like muon decay and a_μ, to the classical parameter $\gamma_{\mu|n}$.

2.4 Helicity Precession of Fermions in Gravitational Fields

The study of the spin evolution of a spin-1/2 particle on a circular orbit in an idealized storage ring requires that expressions for the helicity transition rate be converted into amplitudes and projected into the laboratory frame. Assuming that a beam of spin-1/2 particles follows a circular orbit in the ring, allowing for vertical and horizontal fluctuations about the beam's mean trajectory and by using cylindrical coordinates (r, θ, z, τ) to describe an accelerated frame tangent to the beam orbit, one finds

$$P^1 = -i\hbar \frac{\partial}{\partial r}, \quad P^2 = \frac{-i\hbar}{r}\frac{\partial}{\partial \theta}, \quad P^3 = -i\hbar \frac{\partial}{\partial z}, \tag{2.46}$$

from which one can define the operator $\boldsymbol{R} = \left(0, 0, -\frac{1}{r} P^2\right)$. If the mean orbital radius is r_0 and the orbital frequency is ω_0 in the laboratory frame, then the relationship between the accelerated frame and the laboratory frame in Cartesian coordinates (x, y, z, t), whose origin is at the centre of the storage ring, is given by [46,47]

$$\begin{aligned} x &= (r_0 + \delta r)\cos(\gamma\omega_0\tau) - \gamma\,(r_0\,\delta\theta)\sin(\gamma\omega_0\tau) \\ y &= (r_0 + \delta r)\sin(\gamma\omega_0\tau) + \gamma\,(r_0\,\delta\theta)\cos(\gamma\omega_0\tau) \\ t &= \gamma\left(\tau + r_0^2\,\omega_0\,\delta\theta\right), \end{aligned} \tag{2.47}$$

where δr and $\delta\theta$ are the radial and angular fluctuations about the mean orbit, $\gamma = \left(1 - \omega_0^2 r_0^2\right)^{-1/2}$, and τ is the proper time.

In order to obtain the spin-flip transition amplitude, it is useful to use the Dirac representation and evaluate it in the instantaneous rest frame of the particle. The corresponding ket vectors [48] are

$$|+\rangle_{\rm up} \equiv \begin{pmatrix} \begin{pmatrix} 1 \\ 0 \end{pmatrix} \\ \\ \begin{pmatrix} 0 \\ 0 \end{pmatrix} \end{pmatrix} \qquad |-\rangle_{\rm up} \equiv \begin{pmatrix} \begin{pmatrix} 0 \\ 1 \end{pmatrix} \\ \\ \begin{pmatrix} 0 \\ 0 \end{pmatrix} \end{pmatrix}$$

$$|+\rangle_{\rm dn} \equiv \begin{pmatrix} \begin{pmatrix} 0 \\ 0 \end{pmatrix} \\ \\ \begin{pmatrix} 1 \\ 0 \end{pmatrix} \end{pmatrix} \qquad |-\rangle_{\rm dn} \equiv \begin{pmatrix} \begin{pmatrix} 0 \\ 0 \end{pmatrix} \\ \\ \begin{pmatrix} 0 \\ 1 \end{pmatrix} \end{pmatrix}, \tag{2.48}$$

and the spin-flip transition amplitude of a quantum operator Q is

$$\langle Q \rangle \equiv \langle \mp | Q | \pm \rangle_{\rm up/dn}. \tag{2.49}$$

It follows that some arbitrary spin-independent quantum number $\boldsymbol{K}$ coupled to $\boldsymbol{\sigma}$, obeys the equations

$$\langle (\boldsymbol{\sigma} \cdot \boldsymbol{K}) \rangle = \left(\hat{\boldsymbol{x}}^1 \pm i\, \hat{\boldsymbol{x}}^2 \right) \cdot \boldsymbol{K} \tag{2.50}$$

$$\langle (\boldsymbol{\sigma} \cdot \boldsymbol{K}) \beta \rangle = \pm \left(\hat{\boldsymbol{x}}^1 \pm i\, \hat{\boldsymbol{x}}^2 \right) \cdot \boldsymbol{K}, \tag{2.51}$$

all other matrix elements vanish, and the overall $\pm$ in (2.51) denotes the sign of the contribution specific to the up/dn state, respectively.

One can also verify [49] that the unit vectors $\hat{\boldsymbol{x}}^1(\tau)$ and $\hat{\boldsymbol{x}}^2(\tau)$ are related to the time-independent Cartesian unit vectors $\hat{\boldsymbol{x}}$ and $\hat{\boldsymbol{y}}$ in the laboratory frame by

$$\hat{\boldsymbol{x}}^1(\tau) \pm i\, \hat{\boldsymbol{x}}^2(\tau) = \left(\hat{\boldsymbol{x}} \pm i\, \hat{\boldsymbol{y}} \right) e^{\mp i \gamma \omega_0 \tau} = \left(\hat{\boldsymbol{x}} \pm i\, \hat{\boldsymbol{y}} \right) e^{\mp i \omega_0 t}. \tag{2.52}$$

Using (2.50) and (2.51), the rate of change of the spin-flip transition can be calculated

$$\begin{aligned} \left\langle \frac{dh(t)}{dt} \right\rangle &= \frac{1}{|\boldsymbol{\pi}|} \left(\hat{\boldsymbol{x}}^1(t) \pm i\, \hat{\boldsymbol{x}}^2(t) \right) \cdot \left[\boldsymbol{\Lambda}^0 \pm \boldsymbol{\Lambda}^1 \right] = \\ &= \frac{1}{|\boldsymbol{\pi}|} \left(\hat{\boldsymbol{x}} \pm i\, \hat{\boldsymbol{y}} \right) \cdot \left[\boldsymbol{\Lambda}^0 \pm \boldsymbol{\Lambda}^1 \right] e^{\mp i \omega_0 t}, \end{aligned} \tag{2.53}$$

where $\boldsymbol{\Lambda}^0$ and $\boldsymbol{\Lambda}^1$ are the amplitudes corresponding to (2.50) and (2.51), respectively. The spin evolution is then

$$\begin{aligned} \langle h(t) \rangle &= \frac{1}{|\boldsymbol{\pi}|} \left(\hat{\boldsymbol{x}} \pm i\, \hat{\boldsymbol{y}} \right) \cdot \boldsymbol{\pi} + \int_0^t \left\langle \frac{dh(t')}{dt} \right\rangle dt' \\ &= \frac{1}{|\boldsymbol{\pi}|} \left(\hat{\boldsymbol{x}} \pm i\, \hat{\boldsymbol{y}} \right) \cdot \left[\boldsymbol{\pi} \pm \frac{i}{\omega_0} \left(\boldsymbol{\Lambda}^0 \pm \boldsymbol{\Lambda}^1 \right) \left(e^{\mp i \omega_0 t} - 1 \right) \right]. \end{aligned} \tag{2.54}$$

By evaluating (2.50) and (2.51) for each of the cases under consideration, the amplitudes corresponding to the Dirac Hamiltonian in its original representation, and in its FW- and CT-transformed counterparts, can be obtained (see Appendix D).

One can show that, for the helicity evolution involving the original Dirac Hamiltonian [50]

$$\mathbf{\Lambda}^0_{\text{Dirac}} = \boldsymbol{\omega} \times \boldsymbol{\pi} + \frac{i\hbar}{2}\epsilon^i{}_{jk}\left(\nabla_i \omega^j\right)\hat{\boldsymbol{x}}^k + (\boldsymbol{\omega} \times \boldsymbol{x}) \times \boldsymbol{R} + \epsilon_{klm}\left[\boldsymbol{\nabla}\left(\omega^l x^m\right)\right]\pi^k$$

$$-e\,(\boldsymbol{\omega} \times \boldsymbol{x})^k\,[(\nabla_k A_l) - (\nabla_l A_k)]\,\hat{\boldsymbol{x}}^l - e\,\boldsymbol{\nabla}\varphi - \boldsymbol{\nabla}(\nabla_0 \Phi_{\text{G}}) \tag{2.55}$$

$$\mathbf{\Lambda}^1_{\text{Dirac}} = -m\,\boldsymbol{\nabla}\,(\boldsymbol{a} \cdot \boldsymbol{x}) + \mathbf{\Lambda}^1_\kappa\,, \tag{2.56}$$

$$\mathbf{\Lambda}^1_\kappa = \frac{\kappa e \hbar}{2m}\left[(1 + \boldsymbol{a} \cdot \boldsymbol{x})\left(\frac{2}{\hbar}\boldsymbol{B} \times \boldsymbol{\pi} + i\epsilon^i{}_{jk}\left(\nabla_i B^j\right)\hat{\boldsymbol{x}}^k\right) + i\,\boldsymbol{\nabla}\,(\boldsymbol{a} \cdot \boldsymbol{x}) \times \boldsymbol{B}\right], \tag{2.57}$$

where $\boldsymbol{\omega} = \gamma^2\,\omega_0\,\hat{z}$ [46,51,52] in the rotating frame of reference. In an ideal storage ring, the corrections proportional to $\boldsymbol{a}$ found in (2.57) and several other expressions for Λ given below are of second order because $B = m\omega/e$. This relationship also ensures that Λ^1_κ is mass independent. The remaining terms in (2.57) do not contribute to $g - 2$ experiments because of geometrical constraints or because B is uniform, but would contribute if these constraints were relaxed.

The first and second terms in (2.55) originate from the Mashhoon coupling in the Hamiltonian and the second term would not occur in experiments with ω constant. The first and fourth terms play quite a role in $g - 2$ experiments and are discussed below. The sixth term is well known and can also be found in the calculation of Sakurai [48]. Together, the fifth and sixth terms form something similar to the Lorentz force in a rotating frame.

The third term vanishes in Cartesian coordinates [11], but not in general, when the coordinates are not Cartesian. Its contribution to the precession equation is $\boldsymbol{\sigma} \cdot [(\boldsymbol{\omega} \times \boldsymbol{x}) \times \boldsymbol{R}]$ and has the same dimensions as the term $\boldsymbol{\sigma} \cdot (\boldsymbol{\omega} \times \boldsymbol{p})$ that appears in the Thomas-BMT equation [53]. The third term may be small for the geometry of $g - 2$ experiments, but not so for other types of spin motion like those considered in spin rotators and Siberian snakes.

The last term in (2.55) is new, but can be removed by a gauge transformation in the absence of topological singularities. In the latter case the integral of Φ_G over a closed path encircling the singularity is gauge invariant up to an integer multiple of 2π. Barring this possibility, the calculation based on Eqs. (2.53) and (2.54) implies the conservation of helicity for massless particles, as expected [54].

However, particles are known to acquire an effective mass [55] when acted upon by inertia-gravitation as shown in Sect. (1.3.4).

By choosing Cartesian coordinates so that the term $(\boldsymbol{\omega} \times \boldsymbol{x}) \times \boldsymbol{R}$ vanishes, dropping second-order terms in ω, and combining the remaining first and fourth terms in (2.55) with the term $(\kappa e/m)\,\boldsymbol{B} \times \boldsymbol{\pi}$ in (2.56), one finds

$$\left\langle \frac{dh}{dt} \right\rangle \simeq \frac{1}{|\boldsymbol{\pi}|}\left(\hat{\boldsymbol{x}}^1 \pm i\hat{\boldsymbol{x}}^2\right) \cdot \left[\pm\frac{\kappa e}{m}\,\boldsymbol{B} \times \boldsymbol{\pi} + \epsilon_{ijk}\left(\boldsymbol{\nabla}\omega^i\right)x^j\,\pi^k\right]. \tag{2.58}$$

The Mashhoon term thus disappears irrespective of whether ω is constant or not, as shown in Sect. (2.2.1). The first term proportional to κ and with $B = m\omega/e$ on the right-hand side of (2.58) is the term normally measured in $g-2$ experiments. If ω is inhomogeneous, then the second term also contributes to the helicity precession. Therefore this term can be neglected only for particular geometrical configurations of the parameters involved.

According to [11], it therefore follows that the corresponding expressions for the FW-transformed Hamiltonian are

$$
\begin{aligned}
\mathbf{\Lambda}^0_{\mathrm{FW}} = {} & \mathbf{\Lambda}^0_{\mathrm{Dirac}} + \frac{1}{2m^2}\,\epsilon_{ijk}\left[(\boldsymbol{\omega}\times\boldsymbol{\pi})\times\boldsymbol{\pi} + e\,(2\kappa\,\boldsymbol{E} - \boldsymbol{\nabla}\varphi)\times\boldsymbol{\pi} - \boldsymbol{\nabla}(\nabla_0\Phi_{\mathrm{G}})\times\boldsymbol{\pi}\right]^i \pi^j\,\hat{\boldsymbol{x}}^k \\
& + \frac{\hbar^2}{8m^2}\left[\boldsymbol{\nabla}\left[\boldsymbol{\nabla}\cdot\boldsymbol{\nabla}\left[(\boldsymbol{\omega}\times\boldsymbol{x})\cdot\boldsymbol{\pi}\right] + \left[\nabla_k\,(\boldsymbol{\omega}\times\boldsymbol{\pi})^k\right]\right] + \right. \\
& \quad + e\left[\nabla_k\left(2\kappa\,E^k + \nabla^k\varphi\right)\right] - \boldsymbol{\nabla}\cdot\boldsymbol{\nabla}(\nabla_0\Phi_{\mathrm{G}})\Big] \\
& - \left[(\boldsymbol{\omega}\times\boldsymbol{x})^j\left[\nabla_j\nabla_k\left(\nabla_i\pi^i\right)\right]\hat{\boldsymbol{x}}^k - \boldsymbol{\omega}\times\left[\boldsymbol{\nabla}\left(\nabla_k\pi^k\right)\right]\right]\Big] \\
& + \frac{\hbar^2}{4m^2}\left[(\boldsymbol{\omega}\times\boldsymbol{x})\cdot\boldsymbol{\nabla}\left[\left(\nabla_1 R^3\right)\hat{\boldsymbol{x}}^2 - \left(\nabla_2 R^3\right)\hat{\boldsymbol{x}}^1 - e\left(\left(\nabla^j\nabla_j A_k\right) - \left(\nabla^j\nabla_k A_j\right)\right)\hat{\boldsymbol{x}}^k\right]\right. \\
& + \left[R^3\left[\epsilon^2{}_{jk}\left(\nabla_1\omega^j\right) - \epsilon^1{}_{jk}\left(\nabla_2\omega^j\right)\right] + \epsilon_{ijk}\,e\left(\left(\nabla_m A^i\right) - \left(\nabla^i A_m\right)\right)\left(\nabla^m\omega^j\right)\right]\hat{\boldsymbol{x}}^k \\
& + \omega^i\left[\epsilon^2{}_{ik}\left(\nabla_1 R^3\right) - \epsilon^1{}_{ik}\left(\nabla_2 R^3\right) + \epsilon_{ijk}\,e\left(\left(\nabla^m\nabla_m A^j\right) - \left(\nabla^m\nabla^j A_m\right)\right)\right]\hat{\boldsymbol{x}}^k\Big] \\
& + \frac{i\hbar}{4m^2}\,\epsilon^i{}_{jk}\,\nabla_i\left[(\boldsymbol{\omega}\times\boldsymbol{\pi})\times\boldsymbol{\pi} + e\,(2\kappa\,\boldsymbol{E} - \boldsymbol{\nabla}\varphi)\times\boldsymbol{\pi} - \left[\boldsymbol{\nabla}(\nabla_0\Phi_{\mathrm{G}})\right]\times\boldsymbol{\pi}\right]^j\hat{\boldsymbol{x}}^k \\
& - \frac{i\hbar^2}{2m^2}\left[(\boldsymbol{\omega}\times\boldsymbol{x})\cdot\boldsymbol{\nabla}\left[R^3\left(\pi^1\hat{\boldsymbol{x}}^2 - \pi^2\hat{\boldsymbol{x}}^1\right) + e\left[(\nabla_j A_k) - (\nabla_k A_j)\right]\pi^j\hat{\boldsymbol{x}}^k\right]\right. \\
& + \omega^i\left[R^3\left(\epsilon^2{}_{ik}\,\pi^1 - \epsilon^1{}_{ik}\,\pi^2\right) + \epsilon_{ijk}\,e\left[\left(\nabla_m A^j\right) - \left(\nabla^j A_m\right)\right]\pi^m\right]\hat{\boldsymbol{x}}^k\Big]
\end{aligned}
\tag{2.59}
$$

and

$$
\begin{aligned}
\mathbf{\Lambda}^1_{\mathrm{FW}} = {} & -(1+\boldsymbol{a}\cdot\boldsymbol{x})\left[\frac{1}{2m}\,\boldsymbol{\nabla}\,(\boldsymbol{\pi}\cdot\boldsymbol{\pi}) + \frac{1}{m}\,\boldsymbol{\sigma}\cdot(\boldsymbol{R}'\times\boldsymbol{\pi}) + \frac{i\hbar}{2m}\,\epsilon^i{}_{jk}\left(\nabla_i R'^j\right)\hat{\boldsymbol{x}}^k\right] \\
& - \boldsymbol{\nabla}\,(\boldsymbol{a}\cdot\boldsymbol{x})\left[m + \frac{1}{2m}\,\boldsymbol{\pi}\cdot\boldsymbol{\pi}\right] + \frac{\kappa e}{m}\,\boldsymbol{\sigma}\cdot(\boldsymbol{B}\times\boldsymbol{\pi}) + \frac{i\kappa e\hbar}{2m}\,\epsilon^i{}_{jk}\left(\nabla_i B^j\right)\hat{\boldsymbol{x}}^k \\
& - \frac{\hbar}{2m}\,\epsilon^{ij}{}_k\left(\nabla_i\nabla_j\Phi_{\mathrm{G}}\right)\hat{\boldsymbol{x}}^k - \frac{i\hbar}{2m}\left[\boldsymbol{\nabla}\,(\boldsymbol{a}\cdot\boldsymbol{x})\times\boldsymbol{R}' + \left(\nabla^j\left[\nabla_j\nabla_k - \nabla_k\nabla_j\right]\Phi_{\mathrm{G}}\right)\hat{\boldsymbol{x}}^k\right] \\
& - \frac{1}{2m}\left[(\boldsymbol{\nabla}\,(\boldsymbol{a}\cdot\boldsymbol{x})\times\boldsymbol{\pi})\times\boldsymbol{\pi} + 2\,\boldsymbol{\nabla}(\boldsymbol{\nabla}\Phi_{\mathrm{G}}\cdot\boldsymbol{\pi})\right] - \frac{1}{m}\left(\left[\nabla_j\nabla_k - \nabla_k\nabla_j\right]\Phi_{\mathrm{G}}\right)\pi^j\,\hat{\boldsymbol{x}}^k \\
& - \frac{\hbar^2}{8m}\,\boldsymbol{\nabla}\,(\boldsymbol{\nabla}\cdot\boldsymbol{\nabla}\,(\boldsymbol{a}\cdot\boldsymbol{x})) - \frac{i\hbar}{4m}\,\epsilon^i{}_{jk}\left[\nabla_i\,(\boldsymbol{\nabla}\,(\boldsymbol{a}\cdot\boldsymbol{x})\times\boldsymbol{\pi})^j\right]\hat{\boldsymbol{x}}^k.
\end{aligned}
\tag{2.60}
$$

In a local Cartesian frame, the leading-order contributions to the amplitude in the low-energy approximation, of order $1/m$, come from the second and sixth terms of (2.60), which yield the total magnetic moment term $[(1+\kappa)\,e/m]\,\boldsymbol{\sigma}\cdot(\boldsymbol{B}\times\boldsymbol{\pi})$. Other leading-order contributions are due to the eleventh term in (2.60), which comes from the acceleration-induced spin–orbit coupling term first found by Hehl and Ni [56,57], and also the twelfth and thirteenth terms due to the gravitational energy redshift term found in [57]. As for terms of order $1/m^2$, the leading contributions [57] are due to the second to fourth terms found in (2.59). These are the spin–orbit coupling from the Mashhoon effect, the electric field, and gravitational field, respectively. In addition, the seventh and eighth terms of (2.59) are contributions from the Darwin energy terms due to electromagnetism [48] and gravitation, also found in [57], and by Obukhov [58]. All other terms in (2.59) and (2.60) involving gradients of primarily small quantities are negligible by comparison.

It is shown in [11] that the amplitudes for the CT-transformed Hamiltonian are

$$\begin{aligned}
\mathbf{\Lambda}^0_{\mathrm{CT}} = \mathbf{\Lambda}^0_{\mathrm{Dirac}} + \frac{1}{|\boldsymbol{\pi}|}\frac{\kappa e \hbar}{2m}\left[\nabla\left(\boldsymbol{a}\cdot\boldsymbol{x}\right)\boldsymbol{B}\cdot\boldsymbol{\pi} - i\,\nabla\left(\boldsymbol{a}\cdot\boldsymbol{x}\right)\times\left(\boldsymbol{E}\times\boldsymbol{\pi}\right) + \right.\\
\left. + \left(1+\boldsymbol{a}\cdot\boldsymbol{x}\right)\left[\nabla\left(\boldsymbol{B}\cdot\boldsymbol{\pi}\right) - \frac{2}{\hbar}\left(\boldsymbol{E}\times\boldsymbol{\pi}\right)\times\boldsymbol{\pi} - i\,\epsilon^i{}_{jk}\left[\nabla_i\left(\boldsymbol{E}\times\boldsymbol{\pi}\right)^j\right]\hat{\boldsymbol{x}}^k\right]\right]
\end{aligned} \tag{2.61}$$

and

$$\begin{aligned}
\mathbf{\Lambda}^1_{\mathrm{CT}} &= \mathbf{\Lambda}^1_{\kappa} + \frac{\hbar}{|\boldsymbol{\pi}|^3}\left[\frac{2}{\hbar}\left(\boldsymbol{R}'\times\boldsymbol{\pi}\right) + i\,\epsilon^i{}_{jk}\left(\nabla_i R'^j\right)\hat{\boldsymbol{x}}^k\right]\nabla\Phi_{\mathrm{G}}\cdot\boldsymbol{\pi} - \\
&- \frac{1}{|\boldsymbol{\pi}|}\nabla\left(\nabla\Phi_{\mathrm{G}}\cdot\boldsymbol{\pi}\right) + \frac{i\hbar}{|\boldsymbol{\pi}|^3}\boldsymbol{R}'\times\nabla\left(\nabla\Phi_{\mathrm{G}}\cdot\boldsymbol{\pi}\right) + \\
&+ \frac{\hbar}{2|\boldsymbol{\pi}|^2}\sqrt{|\boldsymbol{\pi}|^2+m^2}\left[\frac{2}{\hbar}\left(\nabla\left(\boldsymbol{a}\cdot\boldsymbol{x}\right)\times\boldsymbol{\pi}\right)\times\boldsymbol{\pi} + i\,\epsilon^i{}_{jk}\left[\nabla_i\left(\nabla\left(\boldsymbol{a}\cdot\boldsymbol{x}\right)\times\boldsymbol{\pi}\right)^j\right]\hat{\boldsymbol{x}}^k\right].
\end{aligned} \tag{2.62}$$

Since the CT-Hamiltonian is an ultra-relativistic approximation of the original Dirac Hamiltonian, one expects that many of the terms in (2.61) and (2.62) will be small relative to the contributions due to the original Dirac Hamiltonian, disregarding those due to inhomogeneous fields. There are, nonetheless, a few terms which should make a meaningful contribution. The fifth term in (2.61), for instance, is a spin–orbit coupling due to the electric field which yields a term $\frac{1}{|\boldsymbol{\pi}|}\left(\kappa e/m\right)\left(\boldsymbol{E}\times\boldsymbol{\pi}\right)\times\boldsymbol{\pi}$. Other contributions of note are the second, fourth, and sixth terms of (2.62), which are due to the ultra-relativistic analogues of the Hehl-Ni spin–orbit coupling and gravitational energy redshift terms found in the low-energy approximation.

2.5 Chirality Precession of Fermions in Gravitational Fields

It is of interest to also study the effect of inertia on the chirality precession of spin-1/2 particles, for comparison with the helicity precession. Given that the γ^5 operator in the chiral representation is

$$\gamma^5 = \begin{pmatrix} 1 & 0 \\ 0 & -1 \end{pmatrix}, \tag{2.63}$$

the projection operators for isolating right- and left-handed states are defined as

$$P_{\mathrm{R}} \equiv \frac{1}{2}\left(1+\gamma^5\right), \qquad P_{\mathrm{L}} \equiv \frac{1}{2}\left(1-\gamma^5\right), \tag{2.64}$$

$$\psi \equiv \begin{pmatrix} \varphi_\mathrm{R} \\ \varphi_\mathrm{L} \end{pmatrix}. \tag{2.65}$$

Equivalently, (2.64) can be written as

$$P_\pm = \frac{1}{2}\,(1 \pm \gamma^5). \tag{2.66}$$

Applying the CT transformation on (2.66) leads to [59]

$$P_{\mathrm{CT}\pm} \approx P_\pm \pm \frac{\omega(q)}{2|\boldsymbol{\pi}|}\,\gamma^5\,\beta\,(\boldsymbol{\alpha}\cdot\boldsymbol{\pi})\,. \tag{2.67}$$

Then, to leading order in q,

$$\dot{P}_{\mathrm{CT}\pm} \approx \pm q\left[-\frac{1}{2}\left(1 - \frac{\hbar}{|\boldsymbol{\pi}|^2}\,\boldsymbol{\sigma}\cdot\boldsymbol{R}'\right)\left\{\frac{i}{\hbar}\,[H_0, h] + \frac{1}{|\boldsymbol{\pi}|}\,(\boldsymbol{\alpha}\cdot\boldsymbol{\pi})\cdot\frac{i}{\hbar}\,[H_0,\beta]\,\beta\,\gamma^5\right\} + \right.$$
$$\left. + \frac{i}{\hbar}\,H_1\,\gamma^5\,\beta\right]\beta\,. \tag{2.68}$$

Explicitly, the chirality transition rate (2.68) is given by [11]

$$\begin{aligned}
\frac{1}{|\boldsymbol{\pi}|}\,(\boldsymbol{\alpha}\cdot\boldsymbol{\pi})\cdot\frac{i}{\hbar}\,[H_0,\beta]\,\beta = {} & \frac{2}{|\boldsymbol{\pi}|^2}\left[\sqrt{|\boldsymbol{\pi}|^2 + m^2}\,\Pi_1 - \frac{q^3}{\sqrt{1+q^2}}\,\frac{\hbar}{2m}\Pi_2\right] \\
& + \frac{2i}{\hbar|\boldsymbol{\pi}|^2}\,(1 + \boldsymbol{a}\cdot\boldsymbol{x})\left\{\sqrt{|\boldsymbol{\pi}|^2 + m^2}\,\left(\boldsymbol{\pi}\cdot\boldsymbol{\pi} + \hbar\,\boldsymbol{\sigma}\cdot\boldsymbol{R}'\right)\right. \\
& - \frac{q^3}{\sqrt{1+q^2}}\,\frac{\hbar}{2m}\left[\boldsymbol{\sigma}\cdot\left[-i\hbar\,\boldsymbol{\nabla}\left(\boldsymbol{R}'\cdot\boldsymbol{\pi}\right) + \left(\boldsymbol{R}'\cdot\boldsymbol{\pi}\right)\boldsymbol{\pi}\right]\right. \\
& \left.\left. + \,\hbar\Pi_3 + \boldsymbol{\sigma}\cdot\left[\left(\boldsymbol{R}'\times\boldsymbol{\pi}\right)\times\boldsymbol{\pi}\right]\right]\right\} \\
& + \frac{2}{|\boldsymbol{\pi}|}\left[(\boldsymbol{\nabla}\cdot\boldsymbol{\nabla}\Phi_\mathrm{G}) + i\,\sigma^k\,\epsilon^{ij}{}_k\left(\nabla_i\,\nabla_j\,\Phi_\mathrm{G}\right)\right] \\
& + \frac{2i}{\hbar|\boldsymbol{\pi}|}\,[\boldsymbol{\nabla}\Phi_\mathrm{G}\cdot\boldsymbol{\pi} - i\,\boldsymbol{\sigma}\cdot[\boldsymbol{\nabla}\Phi_\mathrm{G}\times\boldsymbol{\pi}]]\,,
\end{aligned} \tag{2.69}$$

where

$$\Pi_1 = \boldsymbol{\nabla}\,(\boldsymbol{a}\cdot\boldsymbol{x})\cdot\boldsymbol{\pi} + i\,\boldsymbol{\sigma}\cdot[\boldsymbol{\nabla}\,(\boldsymbol{a}\cdot\boldsymbol{x})\times\boldsymbol{\pi}]\,,$$

$$\Pi_2 = [\boldsymbol{\sigma}\cdot\boldsymbol{\nabla}\,(\boldsymbol{a}\cdot\boldsymbol{x})]\,\boldsymbol{R}'\cdot\boldsymbol{\pi} + i\,\boldsymbol{\nabla}\,(\boldsymbol{a}\cdot\boldsymbol{x})\cdot\left(\boldsymbol{R}'\times\boldsymbol{\pi}\right) - \boldsymbol{\sigma}\cdot\left[\boldsymbol{\nabla}\,(\boldsymbol{a}\cdot\boldsymbol{x})\times\left(\boldsymbol{R}'\times\boldsymbol{\pi}\right)\right]\,,$$

$$\Pi_3 = \left[\nabla_k\left(\boldsymbol{R}'\times\boldsymbol{\pi}\right)^k\right] + i\,\sigma^k\,\epsilon^i{}_{jk}\left[\nabla_i\left(\boldsymbol{R}'\times\boldsymbol{\pi}\right)^j\right]\,.$$

Excluding the contributions from the anomalous magnetic moment terms, it is clear from (2.68) that

$$\dot{P}_{\mathrm{CT}\pm}|_{m=0} = 0, \tag{2.70}$$

and so the chirality is a constant of the motion for massless particles (see Appendix E for the derivation in linearized quantum gravity theory). A comprehensive study of chirality transitions in a Schwarzschild field is presented in [60].

Inertial-gravitational fields affect quantum particles in different ways. They interact with particle spins and give rise to quantum phases that can be measured in principle by interferometric means. In this case Φ_G must be calculated over a closed space–time path that can be obtained, for instance, by comparing the phase of a particle at the final position P_f at the final time t_f with that of an identical particle at the same final point P_f, but at the initial time t_i.

Through the Hamiltonian, the fields can also affect the energy levels and the time evolution of observables. In the latter case, inertial fields change the helicity and chirality of particles in ideal storage rings. This has been studied in some detail in Sects. 2.2–2.5. The results independently confirm that the spin-rotation coupling compensates the much larger contribution that comes from the $g = 2$ part of the magnetic moment of a pure Dirac particle. Without this cancellation, g-2 experiments of high accuracy would be more difficult to perform.

In the more general case of an inhomogeneous $\boldsymbol{\omega}$, the Mashhoon term per se essentially disappears, but a new term $\frac{1}{|\pi|}\epsilon^{i}{}_{jk}\left(\omega_{i,1} + i\,\omega_{i,2}\right)x^{j}\,\pi^{k}$ contributes to the spin-precession. In the lowest approximation, the spin precesses with the same angular frequency ω of the particle itself. The ratio of this new term to ω is $\simeq [(\nabla\omega)/\omega]\,x$. It may be possible to conceive of a physical situation in which this term can be observed, which would extend the present knowledge of rotational inertia.

Problems

2.1 *Show that the spin-rotation coupling is contained in the Berry phases calculated in Sects. (1.3.5), (1.3.7), (1.3.10), and (1.3.12).*

2.2 *Derive Eq. (2.21) and discuss its relation to the possible values of* κ_1 *and* κ_2.

2.3 *Derive Eq. (2.1)*

2.4 *Derive Eq. (2.4) of Sect. 2.2.*

References

1. Mashhoon, B.: Phys. Rev. Lett. **61**, 2639 (1988); Phys. Lett. A **139**, 103 (1989); **143**, 176 (1990); **145**, 147 (1990); Phys. Lett. A **306**, 66 (2002); arXiv:0801.2134v1 [gr-qc] 14 Jan 2008. Phys. Rev. Lett. **68**, 3812 (1992)
2. Ramos, J., Mashhoon, B.: arXiv:gr-qc/0601054v2 8 Mar 2006
3. Silenko, A.J., Teryaev, O.V.: arXiv:gr-qc/0612103v2 27 Sep 2007
4. Hehl, F.W., Ni, W.-T.: Phys. Rev. D **42**, 2045 (1990)
5. Ashby, N.: Liv. Rev. Relat. **6**, 1 (2003)
6. Demirel, B., Sponar, S., Hasegawa, Y.: New J. Phys. **17**, 023065 (2015)

7. Danner, A., Demirel, B., Sponar, S., Hasegawa, Y.: J. Phys. Comm. **3**, 035001 (2019)
8. Mashhoon, B., Obukhov, Y. N.: Phys. Rev. D **88**, 064037 (2013)
9. For some recent results see: Deriglazov, A.A., Pupasov-Maksimov, A.M.: arXiv:1312.6247v2 [hep-th] 11 Jan 2014
10. Mashhoon, B.: arXiv:gr-qc/031030v2 17 Apr 2008; arXiv:gr-qc2102.06433v1 12 Feb 2021
11. Mashhoon, B.: Black Holes: Theory and Observation. In: Hehl, F.W., Kiefer, C., Metzler, C. (eds.)Springer, Berlin (1998)
12. Caianiello, E.R.: Lett. Nuovo Cim. **32**, 65 (1981); **41**, 370 (1984); Riv. Nuovo Cimento **15**(4) (1992); Caianiello, E.R., Vilasi, G.: Lett. Nuovo Cimento **30**, 469 (1981); Caianiello, E.R., de Filippo, S., Marmo, G., Vilasi, G.: Lett. Nuovo Cimento **42**, 70 (1985)
13. Gasperini Astroph. M.: Space Sci. **138**, 387 (1987); Toller, M.: Phys. Lett. B **256**, 215 (1991); Pati, A.K.: Europh. Lett. **18**, 285 (1992); Torrome, R.G.: arXiv:gr-qc/0701091 (2007); Friedman, Y., Gofman, Y.: aeXiv:gr-qc/0509004 (2005); Brandt, H.E.: Found. Phys. Lett. **2**, 39 (1989); Nesterenko, V.V., Feoli, A., Lambiase, G., Scarpetta, G.: Phys. Rev. D **60**, 965001 (1998); Sanchez, N., Veneziano, G.: Nucl. Phys. B **333**, 253 (1990); Gasperini, M.: Gen. Rel. Grav. **24**, 219 (1992); Frolov, V.P., Sanchez, N.: Nucl. Phys. B **349**, 815 (1991); Feoli, A., Lambiase, G., Papini, G., Scarpetta, G.: Phys. Lett. A **263**, 147 (1999); Lambiase, G., Papini, G., Scarpetta, G.: Phys. Lett. A **244**, 349 (1998); Rovelli, C., Vidotto, F.: Phys. Rev. Lett. **111**, 091303 (2013)
14. Foldy, L.L., Wouthuysen, S.A.: Phys. Rev. **78**, 30 (1950)
15. Bonse, U., Wroblewski, T.: Phys. Rev. Lett. **51**, 1401 (1983)
16. Page, L.A.: Phys. Rev. Lett. **35**, 543 (1975)
17. Werner, S.A., Staudenman, J.-L., Colella, R.: Phys. Rev. Lett. **42**, 1103 (1979)
18. Cai, Y.Q., Papini, G.: Phys. Rev. Lett. **66**, 1259 (1991); **68**, 3811 (1992)
19. Bailey, J. et al.: Nucl. Phys. **B150**, 1 (1979)
20. Farley, F.J.M., Picasso, E.: Advanced Series in High-Energy Physics, vol. 7, p. 479. In: Kinoshita, T. (eds.) Quantum electrodynamics. World Scientific, Singapore (1990)
21. Farley, F.J., Picasso, E.: Ann. Rev. Nucl. Part. Sci. **29**, 243 (1979)
22. Papini, G., Lambiase, G.: Phys. Lett. A **94**, 175 (2002)
23. Brown, H.N., et al.: Muon (g-2) collaboration. Phys. Rev. **D62**, 091101 (2000)
24. Brown, H.N., et al.: Muon (g-2) collaboration. Phys. Rev. Lett. **86**, 2227 (2001)
25. Schiff, L.I.: Phys. Rev. Lett. **1**, 254 (1958)
26. Leitner, J., Okubo, S.: Phys. Rev. **136**, B1542 (1964)
27. Hari Dass, N.D.: Phys. Rev. Lett. **36**, 393 (1976); Ann. Phys. (NY) **107**, 337 (1977)
28. Almeida, L.D., Matsas, G.E.A., Natale, A.A.: Phys. Rev. D **39**, 677 (1989)
29. Papini, G.: Advances in the interplay between quantum and gravity physics. In: de Sabbata, V., Zheltukhin, A. (eds.) Kluwer Academic Publishers, Dordrecht (2002); Papini, G., Lambiase, G.: Phys. Lett. A **294**, 175 (2002)
30. Effective non-Hermitean interaction terms are at times considered in general relativity. This in general reflects the fact that in the presence of a gravitational field the number of particles in a certain channel can change. For example, see Wald, R.M.: Phys. Rev. D **21**, 2742 (1980)
31. Kenny, B.G., Sachs, R.G.: Phys. Rev. D **8**, 1605 (1973)
32. Sachs, R.G.: Phys. Rev. D **33**, 3283 (1986)
33. Hayasaka, H., Takeuchi, S.: Phys. Rev. Lett. **63**, 2701 (1989)
34. Faller, J.E., Hollander, W.J., Nelson, P.G., McHugh, M.P.: Phys. Rev. Lett. **64**, 825 (1990)
35. Nitschke, J.M., Wilmarth, P.A.: Phys. Rev. Lett. **64**, 2115 (1990)
36. Gharibyan, V.: Mod. Phys. Lett. A **35**, 2050079 (2020)
37. Muon g-2 Collaboration, Phys. Rev. Lett. **126**, 141801 (2021)
38. Garisto, D.: Symmetry Magazine, 10/20/20
39. Litvinov, Y.A., et al.: Phys. Rev. Lett. **99**, 262501 (2007)
40. Litvinov, Yu.A., et al.: Phys. Lett. B **664**, 162 (2008)
41. Takahashi, R., Nakamura, T.: ApJ **595**, 1039 (2003)
42. Lambiase, G., Papini, G.: Phys. Lett. A **377**, 1021 (2013)
43. Jackson, J.D.: Classical Electrodynamics. Wiley, Inc, New York (1999)

44. Vogel, M., et al.: Nucl. Instrum. Meth. **B** 235, 7 (2005); Moskovkin, D.L., et al.: Phys. Rev. A **70**, 032105 (2004)
45. Blundell, S.A., et al.: 1997 Phys. Rev. A **55**, 1857 (1997)
46. Bell, J.S., Leinaas, J.: Nucl. Phys. B **284**, 488 (1987)
47. Mashhoon, B.: Advances in General Relativity and Cosmology, edited by G. Ferrarese, Pitagora (Bologna) (2003).(gr-qc/0301065)
48. Sakurai, J.J.: Advanced Quantum Mechanics. Addison-Wesley, New York (1967)
49. Mashhoon, B.: Phys. Lett. A **139**, 103 (1989)
50. Singh, D., Mobed, N., Papini, G.: J. Phys. A:Math. Gen. **37**, 8329 (2004)
51. Irvine, W.M.: Physica **30**, 1160 (1964)
52. Mashhoon, B., Obukhov, Y.N.: Phys. Rev. D **88**, 064037 (2013)
53. Montague, B.W.: Phys. Rep. **113**, 1 (1984)
54. Itzykson, C., Zuber, J.-B.: Quantum Field Theory. McGraw-Hill, Toronto (1980)
55. Weyl, H.: Zeits. f. Physik **56**, 330 (1929)
56. Hehl, F.W., Ni, W.-T.: Phys. Rev. D **42**, 2045 (1990)
57. Singh, D., Papini, G.: Nuovo Cimento **115**, 223 (2000)
58. Obukhov, Yu.N.: Fortsch. Phys. **50**, 711 (2002)
59. Chen, C.X., Papini, G., Mobed, N., Lambiase, G., Scarpetta, G.: Nuovo Cimento B **114**, 1335 (1999)
60. Singh, D.: Phys. Rev. D **71**, 105003 (2005)

Interferometers in Gravitational Fields

3

Abstract

Interferometry is the ideal technique to measure phase differences. Superconducting interferometers are also considered and comparisons are made between massive particle interferometers and those using photons. Various metrics are considered.

3.1 Introduction

Non-relativistic and relativistic wave equations in weak, external gravitational and inertial fields have been dealt with in Chap. 1. Only two fundamental aspects of the interaction have been considered: the generation of quantum phases and spin-gravity coupling.

Quantum phases can be calculated exactly to first order in the field and in a manifestly covariant way for Klein–Gordon, Maxwell, Proca, Dirac and spin-2 equations. They can then be tested experimentally.

The behaviour of quantum systems is consistent with that predicted by general relativity, intended as a theory of both gravity and inertia, down to distances $\sim 10^{-8}$ cm. This is inferred from measurements on superconducting electrons ($\sim 10^{-5}$cm) and on neutrons ($\sim 10^{-8}$cm) which are not tests of general relativity *per se*, but confirm that the behaviour of inertia and Newtonian gravity is that predicted by wave equations that satisfy the principle of general covariance. Atomic and molecular interferometers will likely push this limit down to $10^{-9} - 10^{-11}$ cm and lead to new tests of general relativity. Prime candidates in this regard are a correction term to the gravitational redshift of photons and a (general) relativistic correction to the gravitational field of Earth. The Lense–Thirring effect of Earth has been recently observed [1–3].

The Mashhoon effect [4] offers interesting insights into the interaction of inertia-gravity with spin. Spin is, of course, a quantum degree of freedom *par excellence*. Spin-rotation coupling plays a role in precision measurements of the anomalous

G. Lambiase and G. Papini, *The Interaction of Spin with Gravity in Particle Physics*, Lecture Notes in Physics 993, https://doi.org/10.1007/978-3-030-84771-5_3

magnetic moment of muons, thus extending tests of the covariant Dirac equation down to distances comparable with the muon's wavelength, or $\sim 2 \times 10^{-13}$ cm.

Several applications of solutions (1.33) and (1.37) to superconductors, gyroscopy and interferometry can be now discussed.

3.1.1 Superconductors

In the theory of superconductivity by Bardeen, Cooper, and Schrieffer [5] the Schroedinger equation for superconductors in electromagnetic fields is given by

$$i\frac{\partial\psi}{\partial t} = \left\{\left[\frac{1}{2m}\left(\mathbf{p} - e\mathbf{A}\right)^2 - A_0 + V_I\right]\right\}\psi, \tag{3.1}$$

where V_I is the spin-lattice interaction responsible for electron pairing and the electromagnetic potentials $\mathbf{A}$ and A_0 are determined by the Maxwell equations. In stationary conditions the electric field inside the superconductor is $\mathbf{E} = 0$ and the magnetic field is given by

$$\nabla^2\mathbf{A} = \lambda^2\mathbf{A}, \tag{3.2}$$

where λ is a constant, about 10^{-5} cm, for most superconductors. This equation means that the magnetic field inside a superconductor decreases exponentially at distances larger than the penetration depth λ. Equation (3.1) must now be compared with (1.32) that also accounts for the presence of *stationary* gravitational fields. One can immediately draw the following conclusions [6,7]. First, the quantity that must now vanish inside the superconductor is $\overrightarrow{\nabla}(A_0 - \frac{1}{2}\frac{mc^2}{e}\gamma_{00}) = 0$. That is, contrary to the gravity free case, the gravitational field generates the electric field $\mathbf{E} = (mc^2/2e)\nabla\gamma_{00}$ inside the superconductor. If the field is Newtonian, then $\mathbf{E} = \frac{m\mathbf{g}}{e}$ which is the field gravity would produce inside normal conductors [8] when the lattice is sufficiently rigid. Similarly, the magnetic field

$$B_i = -\frac{mc^2}{e}\varepsilon_{ijk}\partial^j\gamma^{0k}$$

is generated well inside the superconductor where the magnetic field would vanish in the gravity free case. Finally, whenever the integration path links a singularity as in a multiply connected superconductor, the total flux

$$\oint(A_i - \frac{mc^2}{e}\gamma_{0i})dx^i = n\frac{\hbar c}{2e}$$

is quantized, rather than just the flux of B_i. This also means that γ_{0i} could be measured if the magnetic field generated was sufficiently large.

When the superconductor rotates $\gamma_{0i} = (\frac{\boldsymbol{\omega}\times\mathbf{r}}{c})_i$, one finds $\mathbf{B} = \frac{2mc}{e}\boldsymbol{\omega}$ [7], which is the London moment of rotating superconductors measured in [9]. This result offers tangible evidence that fictitious forces, due to the introduction of non-inertial frames, interact with a quantum system in ways that are compatible with Einstein's theory down to lengths of order $\leq 10^{-5}$ cm. This was not known before 1966 [7].

These conclusions only apply to *stationary* gravitational fields. Other examples of gravity-induced electric and magnetic fields are discussed in the literature [10].

Gyroscopy is completely controlled by quantum phases. For photons, one finds

$$\frac{(\Delta\chi)_{ph}}{(\Delta\chi)_{part}} = \frac{\lambda_c}{(\lambda)_{ph}},$$

where λ_c is the Compton wavelength of the particle circulating in the interferometer. This ratio indicates that particle interferometers are more sensitive than photon interferometers for particle masses $m > \frac{h\nu_{ph}}{c^2}$.

On applying (1.46) to photons, one finds that the time integral part of ξ yields

$$-\frac{1}{4}\int_P^x dz^0(\gamma_{\alpha 0,\beta} - \gamma_{\beta 0,\alpha})S^{\alpha\beta} - \frac{1}{2}\int_P^x dz^0\gamma_{\alpha\beta,0}T^{\alpha\beta} =$$

$$-\frac{1}{2}\int_P^x dz^0\gamma_{i0,j}S^{ij} = \int dt\omega S_z \tag{3.3}$$

which represents the spin-rotation coupling, or Mashhoon effect, for photons [4, 11].

3.1.2 Gravitational Waves and Superfluids

Assume for simplicity that a square interferometer of side ℓ and vertices A, B, C, D lies in the (xy)-plane and that a gravitational wave travels in the z-direction. It is possible to choose a transverse-traceless gauge, in which [10]

$$\gamma_{xx}^{(1)} = -\gamma_{yy}^{(1)} = f_+ \cos\omega(t - z/c); \quad \gamma_{xy}^{(1)} = \gamma_{yx}^{(1)} = f_\times \cos\omega(t - z/c). \tag{3.4}$$

This metric must now be substituted into (1.37) and the integrals evaluated, taking into account that the interferometer beams are split at A at time t and interfere at C. Setting $f_+ = f_\times$ and using the non-relativistic approximations $k^0 \approx \frac{mc}{\hbar} + \frac{\hbar\kappa^2}{2mc}$, with $\kappa = mv/\hbar$, one obtains

$$\Delta\chi = \Delta\chi_2 + \Delta\chi_1 = \frac{4fv\kappa}{\omega}\left\{\left(1 + \frac{\hbar v\kappa}{2mc^2}\right)\sin\left(\omega t + \frac{\ell\omega}{v}\right)\sin\frac{\ell\omega}{2v} - \right.$$

$$\left. - \frac{\ell\hbar\kappa\omega}{4mc^2}\sin\left(\omega t + \frac{\ell\omega}{2v}\right)\right\}\sin\frac{\ell\omega}{2v}, \tag{3.5}$$

where the first term in curly brackets is $\Delta\chi_2$. For the sake of comparison, when photons of frequency $\frac{\tilde{\omega}}{2\pi}$ are used in an entirely similar setup, the phase difference is

$$\Delta\chi = \Delta\chi_2 = \frac{4f\tilde{\omega}}{\omega}\sin^2\left(\frac{\omega\ell}{2c}\right)\sin\left(\omega t + \frac{\omega\ell}{c}\right). \tag{3.6}$$

The second term in $\Delta\chi_2$ is a correction of order $(\frac{v}{c})^2$. The third term plays a role at high frequencies. If the superfluid loop is in the (xz)-plane, the result can be obtained by applying Eq. (1.37) to a space-like square path of length $2a$ in the x-direction and length $2b$ in the z-direction. A simple calculation performed in the transverse-traceless gauge yields

$$\Delta\chi = -\frac{2m\omega^2 a^2 b}{\hbar c} f_+ \cos\omega t \tag{3.7}$$

in the approximation $a, b << \frac{c}{\omega}$. When the latter restriction is removed, the result is

$$\Delta\chi = -\frac{m\omega a^2}{\hbar} f_+ \sin\omega\left(t - \frac{2b}{c}\right) \tag{3.8}$$

and is more suitable for the high frequency range of the spectrum. The last two equations must be multiplied by a factor N if the superfluid path consists of N loops. These equations yield the result of [12] without the use of particular (Fermi normal) coordinates and the restrictions that they imply.

A measure of unity can be brought to the problem of particle interferometry in weak gravitational fields. This can be accomplished by means of the solution (1.37), valid whenever a solution of the field-free equation is known. This solution, exact to first order in $\gamma_{\mu\nu}$, can then be calculated explicitly by means of path integrals. The contribution of the gravitational field amounts to a phase factor which is obviously amplified by the number of complete paths the particles travel through before interfering. In the particular problem considered, the gravitational field behaves effectively as a vector potential $G_\lambda(x)$. While this was known to occur in similar circumstances for stationary fields, the present result also applies to time dependent gravitational fields and to fully relativistic systems.

The phase shift induced by an incoming gravitational wave provides an example of a detection scheme which is, in principle, independent of any resonance condition. This scheme looks ideally suited for the detection of high frequency gravitational radiation. In this case, however, the response time τ of the antenna becomes critical. A calculation based on the sudden approximation indicates that a change in the state of the system cannot take place before a time $\tau \sim \frac{\hbar}{\Delta E}$ has elapsed, where ΔE is the change in the Hamiltonian induced, in this case by the gravitational perturbation. Since

$$\Delta E \approx \frac{\gamma^{(1)}_{\mu\nu}\phi,^{\mu\nu}\hbar^2}{m} \sim fmv^2\cos\omega t, \tag{3.9}$$

one finds a lower limit

$$\tau \sim \frac{\hbar}{fmv^2} \sim \frac{10^{-13}}{f}. \tag{3.10}$$

The ratio of the third to the first term in (3.5) gives $\sim \frac{\ell v\omega}{4c^2} > 1$ for $\omega > \frac{4c^2}{\ell v} \sim 4 \times 10^{14}s^{-1}$ for $v \sim 10^5$cm/s and $\ell \sim 10^2$ cm. Thus the third term prevails in magnitude at very high frequencies. A comparison with photon interferometers can now be made. At frequencies $\omega < 4 \times 10^{14}s^{-1}$ the ratio of the dominant terms in (3.5) and (3.6) is about $\frac{mv^2}{\hbar\tilde{\omega}} << 1$, so that photon interferometers appear advantageous, as detectors of gravitational radiation, over massive particle interferometers if the particle mass m is at most of the order of a nucleon mass and $v \sim 10^5$ cm/s. When $\omega > 4 \times 10^{14}s^{-1}$ the ratio of the dominant terms is

$$\frac{m\ell v^3\omega}{4\hbar c^2\tilde{\omega}} \sim \frac{1}{4} \times 10^{-4}\frac{\omega}{\tilde{\omega}} \gtreqless 1\,, \quad \text{for } \omega \gtreqless 4 \times 10^4\tilde{\omega}\,. \tag{3.11}$$

Therefore photon interferometers are more advantageous even in the range $4 \times 10^{14}s^{-1} < \omega < 4 \times 10^4\tilde{\omega}$. Only for graviton frequencies $\omega > 4 \times 10^4\tilde{\omega} > 4 \times 10^{14}s^{-1}$ particle interferometers become in principle more sensitive. These considerations apply only to the nature of the particles circulating in the interferometer. Much more than that is required, of course, in a realistic detector of gravitational waves.

The case of superfluid interferometers deserves special consideration. Because of their extreme coherence all particles in the system are described by the same wave function. Since this must be single valued, one has $\chi = 2\pi n$, where n is an integer, whenever the particles path links a multiconnected region. It then follows from (1.23) that the total flux of G_λ plus that of any electromagnetic field present is quantized. This extends to the time-varying regime what was already known to apply to the time-independent and stationary cases [6,13]. If, in particular, the total flux is initially zero and the loop is then exposed to a gravitational field, a magnetic field is generated to compensate for the change in total flux. This renders flux changes in superconductors vastly more measurable over those of neutral superfluids. A large number of superconducting coils does, in principle, increase the phase difference by a corresponding factor.

3.2 Interferometers in Various Metrics

3.2.1 Interferometer in the Field of Earth

The Berry phase due to the field of Earth is calculated first. If Earth is assumed spherical, then its gravitational field may be described by the Schwarzschild metric [14]

$$d\tau^2 = -\left(1 - \frac{2MG}{c^2r}\right)dt^2 + \frac{1}{c^2}\frac{dr^2}{1 - \frac{2MG}{c^2r}} + \frac{r^2}{c^2}d\Theta^2 + \frac{r^2}{c^2}\sin^2\Theta d\phi^2, \tag{3.12}$$

where M represents the mass of Earth. Equation (1.38) is now applied to a square interferometer of vertices A, B, C, D, in which a beam of particles is split at A and the resulting beams interfere at A again after travelling along the opposite paths

ABCDA and ADCBA. Since the particles are assumed to move with constant speed v, the integration over the time portion of the space–time loop may be reduced to space integrations by choosing the limits of integration appropriately. Assuming that the linear dimension of the interferometer ℓ is such that $\frac{\ell}{R} << 1$, where R is Earth's radius, and expanding $\frac{1}{r}$ in the neighbourhood of $\frac{1}{R}$ up to third order, one obtains

$$\begin{aligned}\Delta\chi = & \frac{MG\kappa^{\circ}\, c\ell^2}{c^2R^2}\frac{}{v}(\cos\alpha - \cos\theta) - \frac{3}{2}\frac{MG\kappa^{\circ}\,\ell\; c\ell^2}{c^2R^2\; R\; v}(\cos^2\alpha - \cos^2\theta) + \\ & + \frac{MG\kappa}{c^2R^2}\ell^2\cos\alpha\cos\theta(\cos\alpha - \cos\theta) - \\ & - \frac{2MG\kappa\;\ell}{c^2R^2\; R}\ell^2\cos\alpha\cos\theta(\cos^2\alpha - \cos^2\theta). \end{aligned} \tag{3.13}$$

The parameters are referred to an orthogonal triad of axes x, y, z with origin at A and unit vector $\hat{z}$ directed as the outward normal to the surface of Earth. Side $\bar{AB}$ lies in the (xz)-plane and makes the angle α with $\hat{z}$ and an angle β with side $\bar{AD}$ which lies in the (yz)-plane at an angle θ with $\hat{z}$. The following relations can be used to transform (3.13)

$$\mathbf{A} = \ell^2\sin\beta\hat{N}\,, \quad \mathbf{g} = -\frac{MG}{R^2}\hat{z}\,, \quad \ell = \bar{AB} = \bar{AD}\,, \tag{3.14}$$

where $\hat{N}$ is the normal to the interferometer area A,

$$\hat{N}\cdot\hat{z} = \frac{sin\alpha\sin\theta}{\sin\beta} \equiv \cos\gamma. \tag{3.15}$$

If in addition the particles are non-relativistic so that

$$\kappa^{\circ} \approx \frac{mc}{\hbar} + \frac{\hbar\kappa^2}{2mc} \text{ with } \kappa = \frac{mv}{\hbar} \tag{3.16}$$

one finds

$$\begin{aligned}\Delta\chi = & \{\frac{m^2}{\hbar^2\kappa} + \frac{\kappa}{2c^2}(1 + 2\cos\alpha\cos\theta) - \frac{3}{2}\frac{m^2\ell}{\hbar^2\kappa R}(\cos\alpha + \cos\theta) - \\ & - \frac{\kappa\ell}{c^2R}(\frac{3}{4} + 2\cos\alpha\cos\theta)(\cos\alpha + \cos\theta)\} \mid \mathbf{A}\times\mathbf{g}\mid . \end{aligned} \tag{3.17}$$

The first term coincides with the corresponding term of Ref.[15]. When the interferometer is rotated by π about an axis, e.g. $\bar{AB}$ or $\bar{AD}$, the contribution from this term gives a phase shift

$$\Delta\chi' = \frac{2m^2}{\hbar^2\kappa} \mid \mathbf{A}\times\mathbf{g}\mid = \frac{4\pi m^2 Ag\lambda}{h^2}\sin\gamma \tag{3.18}$$

which is just what has been observed in the COW experiment [16–18]. The first part of the second term also agrees with the result of [15] except for a factor $\frac{1}{2}$, there missing because of the incorrect expansion of κ° used. The remaining terms are new. The ratios of the various terms in (3.17) are $1 : \frac{v^2}{c^2} : \frac{\ell}{R} : \frac{v^2\ell}{c^2 R}$. When $v = 10^{-5}c$, $\ell = 1$ cm, $R = 7 \times 10^8$ cm, one obtains $1 : 10^{-10} : 10^{-6} : 10^{-16}$. The last term in (3.17) is the smallest one and will be dropped. Of the remaining terms, the third one represents a general relativistic effect which is larger than the special-relativistic correction represented by the second term.

Sensitive tests of universality of free fall with ^{87}Rb atoms in opposite spin orientations have been carried out [19,20]. The observed Eötvös ratio is

$$\frac{2(g_- - g_+)}{g_- + g_+} = (0.2 \pm 1.2) \times 10^{-7} \,. \tag{3.19}$$

These measurements also have a bearing on studies of space–times with torsion discussed in Chap. 7.

3.2.2 Rotation

Consider the phase due to the rotation of a frame of reference which can be described, from the point of view of a co-rotating observer, by the metric [21]

$$d\tau^2 = -\left(1 - \frac{\omega^2 r^2}{c^2}\right) dt^2 + 2\frac{\omega}{c^2}(-ydx + xdy)dt + \frac{1}{c^2}(dx^2 + dy^2 + dz^2), \tag{3.20}$$

where $r^2 \equiv x^2 + y^2$ and ω is the angular velocity about the z-axis. The interferometer of the previous subsection is now rotated, but interference occurs at vertex C opposite to A. Eliminating again the time integration and choosing appropriate integration limits along the space part of the loop, one arrives at the result

$$\Delta\chi = -\frac{\omega\ell^2}{c}\left(\kappa^\circ + \kappa\frac{c}{v}\right) \sin\alpha \sin\theta, \tag{3.21}$$

which, on using Eqs. (3.14)–(3.16), becomes

$$\Delta\chi = -\frac{2m\boldsymbol{\omega}\cdot\mathbf{A}}{\hbar} - \frac{\hbar\kappa^2\boldsymbol{\omega}\cdot\mathbf{A}}{2mc^2}\,, \qquad \boldsymbol{\omega} \equiv \omega\hat{z}\,. \tag{3.22}$$

The first term of (3.22) agrees with the result of other non-relativistic and relativistic approaches [8,10,11,13–22] and the second term represents a relativistic correction. When ω is the angular velocity of rotation of Earth, the first term agrees with the experimental result of [18]. For the particular case of the Sagnac effect, beam splitting and interference take place at A. Then

$$\Delta\chi' = -\frac{2\omega\ell^2}{c}\left(\kappa^\circ + \kappa\frac{c}{v}\right)\sin\alpha\,\sin\theta, \tag{3.23}$$

which, with the usual approximations, yields

$$\Delta\chi' = -\frac{4\,m\boldsymbol{\omega}\cdot\mathbf{A}}{\hbar} - \frac{\hbar\kappa^2\boldsymbol{\omega}\cdot\mathbf{A}}{mc^2}. \tag{3.24}$$

The experimental results, where available [11,23], are in excellent agreement with the present calculations. Gyroscopy is therefore controlled by the quantum phase (3.24).

On applying (1.46) to photons, one finds that the time integral part of ξ yields

$$-\frac{1}{4}\int_P^x dz^0(\gamma_{\alpha 0,\beta} - \gamma_{\beta 0,\alpha})S^{\alpha\beta} - \frac{1}{2}\int_P^x dz^0\gamma_{\alpha\beta,0}T^{\alpha\beta} =$$

$$-\frac{1}{2}\int_P^x dz^0\gamma_{i0,j}S^{ij} = \int dt\omega S_z \tag{3.25}$$

which represents the spin-rotation coupling, or Mashhoon effect, for photons [4,11].

Finally, the successful treatment of the rotational case indicates that the gravitational Berry phase introduced provides the appropriate theoretical basis for both optical and non-optical gyroscopy.

3.2.3 The Lense–Thirring Effect for Quantum Systems

The application of Eqs. (3.13) and (3.14) to particle interferometry requires knowledge of the particle paths in space–time and of the field $\gamma_{\mu\nu}$. The latter is well known for the problem of a rotating, homogeneous, solid sphere (of radius a) originally studied by Lense and Thirring [22].

All other parameters being the same, the effect is maximized by taking $R = a$. The result can be further simplified by placing the interferometer in the (xy)-plane and using coordinates (x, y, z) parallel to $(x'.y', z')$ the coordinates of a point on the sphere. If the particles in the beam have mass m and speed v, then

$$k^0 = \frac{mc}{\hbar}\left(1 + \frac{v^2}{2c^2}\right); \qquad k = \frac{mv}{\hbar} \tag{3.26}$$

and one finds the expression

$$\Delta\chi\phi = \frac{4MGm\ell}{5ca\hbar}\left(\frac{\omega\ell}{c}\right)\left[2 - \frac{3}{a^2}(x'^2 + y'^2)\right] \tag{3.27}$$

$$+\frac{MGm\ell}{ca\hbar}\left[\frac{c}{v}\left(\frac{\ell}{a^2}(x' - y') - \frac{3\ell^2}{2a^4}(x'^2 - y'^2)\right) + \frac{3v}{2c}\frac{\ell}{a^2}(x' - y')\right]$$

that represents the phase measured by an observer comoving with the interferometer relative to which the sphere generating the field is spinning about the z'-axis. The first term in (3.27) depends on ω and represents the Lense–Thirring effect of a quantum system. It has a maximum when the interferometer is placed in the neighbourhood of the poles of the sphere ($x' = y' = 0$). The remaining three terms represent gravitational effects that are present even when $\omega = 0$. Similar terms have been calculated for Earth in [10] using a spherical symmetry. The cylindrical symmetry used in this work seems better suited to the problem. These terms vanish when the beam source location is at $x' = y'$ and, in particular at $x' = y' = 0$, at which positions the only contribution to the particle phase shift is that of the Lense–Thirring field. For Earth the first term can also be written as

$$\Delta\chi_{LT}\phi = \frac{2G}{c^2 R_\oplus^5} J_\oplus \frac{m\ell^2}{\hbar}[2R_\oplus^2 - 3(x'^2 + y'^2)], \tag{3.28}$$

where $J_\oplus = \frac{2M_\oplus R_\oplus^2 \omega}{5}$ is the angular momentum of Earth (assumed spherical and homogeneous) and $R_\oplus$ its radius. It is interesting to observe that $\Omega = \frac{G}{2c^2 R_\oplus^3} J_\oplus$ coincides with the effective Lense–Thirring precession frequency of a gyroscope [23,24]. The absolute values of the ratios of the other terms in (3.27) to the first one for $\ell \ll a$, $x' = 0$, $y' = a = R_\oplus$ are respectively

$$\frac{\Delta\chi\phi_2}{\Delta\chi_{LT}\phi} = \frac{5c^2}{4v\omega R_\oplus}, \tag{3.29}$$

$$\frac{\Delta\chi\phi_3}{\Delta\chi_{LT}\phi} = \frac{15c^2\ell}{8v\omega R_\oplus a}, \tag{3.30}$$

$$\frac{\Delta\chi\phi_4}{\Delta\chi_{LT}\phi} = \frac{15c^2}{48v\omega R_\oplus}. \tag{3.31}$$

For neutron interferometers with $\ell \sim 3.5$ cm, $v \sim 10^5$ cm/s and (3.29), (3.30), and (3.31) yield respectively 2.4×10^{12}, 2.0×10^4 and 40.0. For the same values of the parameters, (3.28) gives $|\Delta\chi_{LT}\phi| \sim 8.0 \times 10^{-15}$ rad. While a measurement of $\Delta\chi\phi_2$ and $\Delta\chi\phi_3$ appears feasible, that of the Lense–Thirring effect will require the use of much larger ($\ell \sim 10^2$ cm) neutron interferometers [25].

For the experimental setup with $R = a$ and $\ell \ll a$, one obtains

$$\begin{aligned}\Delta\chi\phi = {} & \frac{4MGm\ell}{5ca\hbar}\left(\frac{\omega\ell}{c}\right)\left(2 - 3\frac{(x'^2 + y'^2)}{a^2}\right) + \\ & + \frac{MGm\ell}{ca\hbar}\left[\frac{c}{v}\left(\frac{2\ell}{a^2}(x' + y') + \frac{5\ell^2}{3a^2} - \frac{5\ell^2}{2a^4}(x'^2 + y'^2)\right)\right. \\ & \left. + \frac{v}{c}\left(-4 + \frac{3\ell}{a^2}(x' + y') + \frac{15\ell^2}{6a^2} - \frac{15\ell^2}{4a^4}(x'^2 + y'^2)\right)\right].\end{aligned} \tag{3.32}$$

The first term of (3.32) coincides with $\Delta\chi_{LT}\phi$ of (3.27) and with (3.28), when applied to Earth. It was first calculated for a superconducting loop [26] using a low velocity approximation. Removing this restriction by using a relativistic wave equation, rather than the Schroedinger equation, introduces into the Lense–Thirring effect terms $\mathcal{O}(\frac{v^2}{c^2})$ omitted from (3.32) for simplicity. It was also noted in [26] that (3.32) should be multiplied by N if the super electrons travelled around N loops before interfering. Despite the larger neutron mass, a superconducting interferometer could therefore compete favourably in similar measurements if both ℓ and N were large. The remaining terms in (3.32) are analogous to $\Delta\chi\phi_2$, $\Delta\chi\phi_3$ and $\Delta\chi\phi_4$ already discussed. They are all larger than the Lense–Thirring effect. An attempt is being made to measure the relativistic frame dragging and geodetic precession of electron spin using ferromagnetic gyroscopes orbiting Earth [27].

3.3 Wave Optics

The applications of Berry phase theory to interferometry given in [28–30] cover a variety of metrics and physical situations, from the study of rotation and Earth's field to the detection of gravitational radiation and the calculation of effects due to a Lense–Thirring background. The same equation and some of its generalizations can also be applied to the study of quantum fluids and boson condensates. Equation (1.93) is applied in this section to the investigation of wave effects in lensing.

Consider, for example, the propagation of gravitons, or of massive, extreme relativistic spin-2 particles in the lensing metric (1.96). Wave optics effects can best be seen by considering a double slit experiment. For simplicity, a planar arrangement is considered in which particles, source, gravitational lens, and observer lie in the same plane. The particles are emitted at S and interfere at O, where the observer is located, following the paths SLO and SPO (see Fig. 3.1).

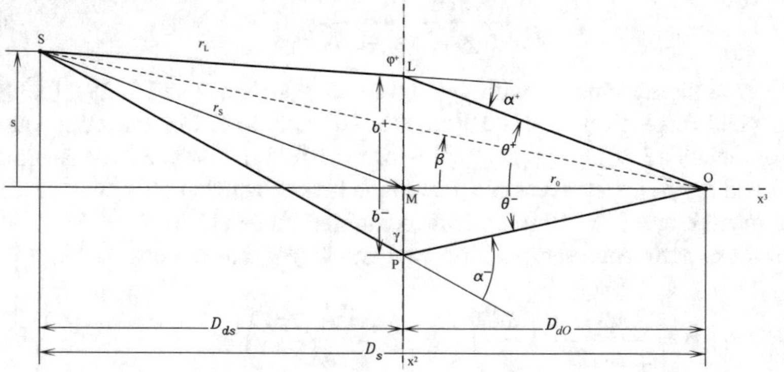

Fig. 3.1 Geometry of a two-image gravitational lens. The spherically symmetric gravitational lens is at M. The solid lines represent particle paths between particle source S and observer O. Particle paths and S, M, O lie in the same plane

The interference and diffraction effects depend on the phase difference experienced by the particles along the different paths and on the gravitational background generated by the spherically symmetric lens at M. M is placed along the diagonal PL a distance b^+ from L and b^- from P. The diagonal PL is along the x_2-axis and MO is along x_3. The distances $r_0 = MS, r_s = MS$ and $s = SH$, the perpendicular distance of S from OM, are the physical variables. The lensing variables are $D_{ds} = MH, D_{d0} = MO, D_s = D_{ds} + D_{d0} = HO$ and the deflection angles are $\theta^+ = L\hat{O}M, \theta^- = M\hat{O}P, \beta = S\hat{O}M, \alpha^+ = \pi - S\hat{L}O, \alpha^- = \pi - S\hat{P}O, \gamma = S\hat{P}L$, $\varphi^+ = \pi + \alpha^+$. A plane wave solution of (1.87) of the form $\phi_{\mu\nu} = e^{-ik_\sigma x^\sigma}\epsilon_{\mu\nu}$ can be used and it is also assumed, for simplicity, that $k^1 = 0$, so that propagation is entirely in the (x^2, x^3)-plane. In this planar setup γ_{11} plays no role. The corresponding wave amplitude is therefore

$$\Phi^0_{\mu\nu} = -ie^{-ik_\sigma x^\sigma}\left\{1 - \frac{1}{2}\left[\int dz^0\gamma_{00,2}(x^0 - z^0)k^2 + \int dz^0\gamma_{00,3}(x^0 - z^0)k^3 - \right.\right.$$
$$- \int dz^0\gamma_{00,2}(x^2 - z^2)k^0 - \int dz^0\gamma_{00,3}(x^3 - z^3)k^0 +$$
$$+ \int dz^2\gamma_{22,3}(x^2 - z^2)k^3 + \int dz^3\gamma_{33,2}(x^3 - z^3)k^2 -$$
$$\left. - \int dz^2\gamma_{22,3}(x^3 - z^3)k^2 - \int dz^3\gamma_{33,2}(x^2 - z^2)k^3\right] +$$
$$\left. + \frac{1}{2}\left[\int dz^0\gamma_{00}k^0 + \int dz^2\gamma_{22}k^2 + \int dz^3\gamma_{33}k^3\right]\right\}. \tag{3.33}$$

The phase must now be calculated along the different paths $SP + PO$ and $SL + LO$, taking into account the values of k^i in the various intervals.

It is convenient to transform all space integrations into integrations over z^0. Along SL one has

$$\varphi = \frac{-GM}{q_{SL}(z^0)^{1/2}}, \quad q_{SL}(z^0) \equiv (r_L - z^0)^2 + b^{+2} + 2(r_L - z^0)b^+\cos\varphi^+, \tag{3.34}$$

and

$$k^2 = k\cos\varphi^+, \quad k^3 = k\sin\varphi^+, \quad r_L\sin\varphi^+ = D_{dS},$$

so that one finds

$$-\frac{\Delta\tilde{\phi}_{SL}}{GM} = 2\int_0^{r_L} dz^0 q_{SL}(z^0)^{-3/2}(z^0 - r_L - b^+\cos\varphi^+)(r_L - z^0)[2k]$$
$$+ \int_0^{r_L} dz^0 q_{SL}(z^0)^{-1/2}[-2k]. \tag{3.35}$$

Analogously, for LO one finds

$$\varphi = -\frac{GM}{q_{LO}(z^0)^{1/2}}, \quad q_{LO}(z^0) \equiv (R_1 - z^0 + r_L)^2 + r_0^2 - 2(R_1 - z^0 + r_L)r_0\cos\theta^+, \tag{3.36}$$

while

$$k^2 = k\sin\theta^+, \quad k^3 = k\cos\theta^+, \quad R_1 = \sqrt{r_0^2 + b^{+2}}\,,$$

and the change in phase is

$$-\frac{\Delta\tilde{\phi}_{LO}}{GM} = 2\int_{r_L}^{r_L+R_1} dz^0 q_{LO}(z^0)^{-3/2}(R_1 - z^0 + r_L + r_0\cos\theta^+)(r_L + R_1 - z^0)[2k] + \int_{r_L}^{r_L+R_1} dz^0 q_{LO}^{-1/2}[-2k]\,. \tag{3.37}$$

For SP one gets

$$\varphi = -\frac{GM}{q_{SP}(z^0)^{1/2}}, \quad q_{SP}(z^0) \equiv b^{-2} + (R - z^0)^2 - 2(R - z^0)b^- \cos\gamma\,, \tag{3.38}$$

$$k^2 = k\cos\gamma, \quad k^3 = k\sin\gamma\,, \quad \tan\gamma = D_{dS}/(s + b^-), \quad R = \sqrt{D_{dS}^2 + (s + b^-)^2}\,,$$

and the corresponding change in phase is

$$-\frac{\Delta\tilde{\phi}_{SP}}{GM} = 2\int_0^R dz^0 q_{SP}(z^0)^{-3/2}(R - z^0 - b^-\cos\gamma)(R - z^0)[2k] + \int_0^R dz^0 q_{SP}(z^0)^{-1/2}[-2k]\,. \tag{3.39}$$

Finally, for PO one obtains

$$\varphi = -\frac{GM}{q_{PO}(z^0)^{1/2}}, \quad q_{PO}(z^0) \equiv r_0^2 + (R_2 + R - z^0)^2 - 2(R_2 + R - z^0)r_0\cos\theta^-\,, \tag{3.40}$$

$$k^2 = -k\sin\theta^-, \quad k^3 = k\cos\theta^-\,, \quad R_2 = \sqrt{r_0^2 + b^{-2}}\,,$$

and the relative change in phase is

$$-\frac{\Delta\tilde{\phi}_{PO}}{GM} = 2\int_R^{R+R_2} dz^0 q_{PO}(z^0)^{-3/2}(z^0 - R_2 - R + r_0\cos\theta^-)(R + R_2 - z^0)[2k] + \int_R^{R+R_2} dz^0 q_{PO}^{-1/2}[-2k]\,. \tag{3.41}$$

The total change in phase therefore is

$$\Delta\tilde{\phi} = \Delta\tilde{\phi}_{SL} + \Delta\tilde{\phi}_{LO} - \Delta\tilde{\phi}_{SP} - \Delta\tilde{\phi}_{PO}\,.$$

All integrations in (3.35), (3.37), (3.39) and (3.41) can be performed exactly. All results can be expressed in terms of physical variables r_s, r_0, b^+, b^-, and s and lensing variables D_s, D_{ds}, D_d, θ^+, θ^-, and β. The final result is

$$\begin{aligned}\Delta\tilde{\phi} = \tilde{y}\Bigg\{&\ln\left(-\sqrt{D_{ds}^2 + \left(s+b^-\right)^2} + b^-\cos\gamma + r_S\right) - \ln\left(b^-\left(1+\cos\gamma\right)\right)\\ &+ \ln\left[b^+\left(1-\cos\varphi^+\right)\right] - \ln\left(r_S - r_L - b^+\cos\varphi^+\right)\\ &+ \ln\left(b^- + r_0\cos\theta^- - \sqrt{b^{-2} + r_0^2}\right) - \ln\left(r_0\left(1+\cos\theta^-\right)\right)\\ &+ \ln\left[r_0\left(1+\cos\theta^+\right)\right] - \ln\left(b^+ + r_0\cos\theta^+ - \sqrt{b^{+2}+r_0^2}\right)\Bigg\}, \end{aligned} \tag{3.42}$$

where

$$\begin{aligned} r_S^2 &= b^{+2} + r_L^2 + 2b^+ r_L \cos\varphi^+ ,\\ r_L^2 &= D_{ds}^2 + (s-b^+)^2 ,\\ \varphi^+ &+ \alpha^+ + \alpha^- + \gamma - \theta^+ - \theta^- = \pi ,\\ \tilde{y} &= 2GMk . \end{aligned}$$

This result is exact to leading order in $\gamma_{\mu\nu}$ and independent of the value of $\tilde{y}$.

A simple quantum mechanical calculation indicates that the probability of finding particles at O contains an oscillating term (two-image interference) that is proportional to $\cos^2 \Delta\tilde{\phi}/2$. In the particular case $b^+ \sim b^- \equiv b\,, \varphi^+ \sim \gamma\,, r_S \sim r_L \gg (b,s)\,, \theta^+ \sim \theta^- \equiv \theta\,, \alpha^+ \sim \alpha^- \equiv \alpha$, one obtains from (3.42) the expression

$$\Delta\tilde{\phi} \sim \tilde{y}\ln\frac{r_L\left(1+\sin\left(\theta-\alpha\right)\right)}{b\sin\left(\theta-\alpha\right)}, \tag{3.43}$$

which is approximate to terms of $O(b/r_L)$ and higher in the argument of the logarithm. The overall probability P_0 of finding particles at O is therefore

$$P_0 \propto \cos^2\left\{\frac{\tilde{y}}{2}\ln\left[\frac{r_L}{2b}\frac{\left(1+\tan\frac{\theta-\alpha}{2}\right)^2}{\tan\frac{\theta-\alpha}{2}}\right]\right\}, \tag{3.44}$$

which exhibits an oscillatory behaviour typical of combined interference and diffraction effects. Higher order terms in b/r_L would in general prevent the logarithmic term from diverging when $\theta \to \alpha$.

The spin-gravity coupling and Mashhoon's helicity-rotation interaction follow from the gravity-induced phase, (1.90). The origin of (1.90) resides in the skew-symmetric part of the space–time connection terms in (1.86), while, in the case of fermions, it is the spinorial connection [11] that accounts for $S_{\alpha\beta}$. The spin term $S_{\alpha\beta}$ affects the polarization of a gravitational wave. It also plays a role in the collision of two gravitational waves. If these are represented by

$$\phi_{22} = -\phi_{33} = \varepsilon_{22} e^{ik(t-x)}, \quad \phi_{23} = \varepsilon_{23} e^{ik(t-x)}$$

and the gravitational background is a wave of the same polarization, but proceeding along the negative direction of the x-axis, then the corresponding metric is

$$ds^2 = 2\Phi'_{02}dx^0dx^2 + 2\Phi'_{03}dx^0dx^3 + 2\Phi'_{12}dx^1dx^2 + 2\Phi_{13}dx^1dx^3 + \\ +2\left(\Phi^0_{23} + \Phi'_{23}\right)dx^2dx^3\,, \tag{3.45}$$

where

$$\begin{aligned}
\Phi'_{02} &= \frac{-ik}{2}\left[(\gamma_{22}\phi_{22} + \gamma_{32}\phi_{23})\,x^2 + (\gamma_{23}\phi_{23} + \gamma_{33}\phi_{23})\,x^3\right]; \qquad (3.46)\\
\Phi'_{03} &= \frac{-ik}{2}\left[(\gamma_{22}\phi_{32} + \gamma_{32}\phi_{23})\,x^2 + (\gamma_{23}\phi_{32} + \gamma_{33}\phi_{33})\,x^3\right];\\
\Phi'_{12} &= \frac{-ik}{2}\left[(\gamma_{22}\phi_{22} + \gamma_{32}\phi_{23})\,x^2 + (\gamma_{23}\phi_{22} + \gamma_{33}\phi_{23})\,x^3\right];\\
\Phi'_{13} &= \frac{-ik}{2}\left[(\gamma_{22}\phi_{23} + \gamma_{32}\phi_{33})\,x^2 + (\gamma_{23}\phi_{32} + \gamma_{33}\phi_{33})\,x^3\right];\\
\Phi'_{23} &= \phi_{22}\,(\gamma_{32} - \gamma_{33})\,,
\end{aligned}$$

and the collision takes place at the origin of the coordinates. The metric (3.45) has a singularity at $x^2 = x^3 = 0$. More complete treatments of this problem show that this is a curvature singularity [31–34].

The geometrical optics of the particles has already been derived. Their deflection is that predicted by Einstein (see Fig. 3.2). The gravitational background behaves as a material medium of index of refraction n given by (1.108). An interesting discussion of geometrical optics in gravitational fields using bilocal operators can be found in [35].

In gravitational lensing [36,37] and in the gravitational lensing of gravitational waves [38], wave effects for a point source depend on the parameter $\tilde{y}$ which gives an indication of the maximum magnification of the wave flux, or, alternatively, of the number of Fresnel zones contributing to lensing. Different values of $\tilde{y}$ require, in general, different approximations, or different solutions of the wave equation. In particular, diffraction effects are expected to be considerable when $\tilde{y} \simeq 1$. In the approach followed, (3.42) holds true regardless of the value of $\tilde{y}$. Wave optics problems usually deal with spherical wave solutions of Helmholtz equation in which gravity appears in the form of a potential. In the present approach, however, gravity makes itself felt in a rather more subtle way than just through a single potential, as evidenced by (1.85).

The framework developed can also be used in the interferometry of atoms and molecules. A laboratory instrument capable of using coherent beams of atoms or molecules would go a long way in probing the interface between gravitational theories and quantum mechanics. For instance, the phase shift of a particle beam in the Lense–Thirring field of Earth is [39]

$$\Delta\tilde{\phi}_{LT} = \frac{4G}{R_\oplus^3}J_\oplus m\ell^2\,, \tag{3.47}$$

Fig. 3.2 Gravitational lensing. A light ring is produced (Einstein ring) when light source, deflecting gravitational object, and observer are aligned. Image credit: ESA/Hubble & NASA

where $J_{\oplus} = 2M_{\oplus}R_{\oplus}^2\Omega/5$ is the angular momentum of Earth (assumed spherical and homogeneous), $R_{\oplus}$ its radius and ℓ the typical dimension of the interferometer. For neutron interferometers with $\ell \sim 10^2$ cm, one finds $\Delta\tilde{\phi}_{LT} \sim 10^{-7}$ rad. The value of the phase difference increases with m and ℓ^2. This suggests that the development and use of large, heavy particle interferometers would be particularly advantageous in attempts to measure gravitational effects. When (3.42) and (3.45) are used in the case of a square interferometer and extreme relativistic particles, one however obtains $\tilde{\phi} \simeq GMk$, irrespective of the size of the interferometer. This is as expected. In fact, for the particular configuration selected, the gravitational flux of the source is completely contained in the integration path $SLOPS$ and $\Delta\tilde{\phi}$ cannot be made larger by increasing the dimensions of the interferometer.

Problems

3.1 *Calculate the contribution of terms containing $S_{\alpha\beta}$ in (1.46) for the metric (3.20) (Skrotskii effect).*

3.2 *Calculate the magnetic field induced by the metric (3.20), that represents rotation of a solid sphere, in a relativistic superconductor using the solution (1.37)-(1.38) of the Klein–Gordon equation.*

3.3 *Use Eqs. (1.37)–(1.38) to calculate the Pound–Rebka effect of Sect.(1.3.3)).*

References

1. Everitt, C.W.F., DeBra, D.B., Parkinson, B.W., Turneaure, J.P., Conklin, J.W., Heifetz, M.I., Keiser, G.M., Silbergleit, A.S., Holmes, T., Kolodziejczak, J., et al.: Phys. Rev. Lett. **106**, 221101 (2011)
2. Iorio, L., Ruggiero, M.L., Corda, C.: Acta Astronaut. **91**, 141 (2013)
3. Iorio, L., Lichtenegger, H.I.M., Ruggiero, M.L., Corda, C.: Astrophys. Space Sci. **331**, 351 (2011)
4. Mashhoon, B.: Phys. Rev. Lett. **61**, 2639 (1988); Phys. Lett. A **139**, 103 (1989); **143**, 176 (1990); **145**, 147 (1990); Phys. Rev. Lett. **68**, 3812 (1992)
5. Bardeen, J., Cooper, L.N., Schrieffer, J.R.: Phys. Rev. **108**, 1175 (1057)
6. DeWitt, B.S.: Phys. Rev. Lett. **16**, 1092 (1966)
7. Papini, G.: Nuovo Cimento **45**, 66 (1966); Phys. Lett. **23**, 418 (1966); Nuovo Cimento B **52**, 136, (1967); Phys. Lett. A **24**, 32 (1967); Nuovo Cimento B **63**, 549 (1969)
8. Schiff, L.I., Barnhill, M.V.: Phys. Rev. **151**, 1067 (1966)
9. Hildebrandt, A.F., Saffren, M.M.: Proc. 9th Int. Conf. on Low-Temp. Phys. Pt. A 459; Hendricks, J.B., Rorschach, H.E., Jr.: ibid. 466; Bol, M., Fairbank, M.M.: ibid. 471; Zimmerman, J.E., Mercereau, J.E.: Phys. Rev. Lett. **14**, 887 (1965)
10. Cai, Y.Q., Papini, G.: Class. Quantum Grav. **6**, 407 (1989)
11. Cai, Y.Q., Papini, G.: Phys. Rev. Lett. **66**, 1259 (1991); **68**, 3811 (1992)
12. Anandan, J., Chiao, R.Y.: Gen. Rel. Grav. **14**, 515 (1982); Anandan, J.: Phys. Rev **D24**, 338 (1984)
13. Papini, G.: Nuovo Cimento **45 B**, 66 (1966); Phys. Lett. **23**, 418 (1966); **A24**, 32 (1967); **A26**, 589 (1968); Nuovo Cimento B **63**, 549 (1969); Lett. Nuovo Cimento **4**, 663(1970); **4**, 1027(1970); Leung, M.-C., Papini, G., Rystephanick, R.G.: Can. J. Phys. **49**, 2754(1971); Rothen, F.: Helv. Phys. Acta **41**, 591(1968); Cabrera, B., Gutfreund, H., Little, W.A.: Phys. Rev. **B25**, 6644 (1982); Meyer, V., Salier, N.: Theoret. Math. Phys. **38**, 270 (1979); Anandan, J.: Phys. Rev. Lett. **47**, 463(1981); Chiao, R.Y.: Phys. Rev. **B 25**, 1655 (1982)
14. Fang, L.Z., Ruffini, R.: Concepts in Relativistic Astrophysics. World Scientific, Singapore (1983)
15. Anandan, J.: Phys. Rev. D **15**, 1448 (1977)
16. Staudenman, J.-L., Werner, S.A., Colella, R., Overhauser, A.W.: Phys. Rev. A **21**, 1419 (1980)
17. Colella, R., Overhauser, A.W., Werner, S.A.: Phys. Rev. Lett. **34**, 1472 (1975)
18. Werner, S.A., Staudenman, J.-L., Colella, R.: Phys. Rev. Lett. **42**, 1103 (1979)
19. Salvi, L., Poli, N., Voletic, V., Tino, G.M.: Phys. Rev. Lett. **120**, 033601 (2018); Liang, L., Poli, N., Salvi, L., Tino, G.M.: Phys. Rev. Lett. **119**, 263601 (2017); Rosi, G., Cacciapuoti, L., Sorrentino, F., Menchetti, M., Prevedelli, M., Tino, G.M.: Phys. Rev. Lett. **114**, 013001 (2015)
20. Duan, X.-C., Den, X.-B., Zhou, M.-K., Ke-Zhang, W.-L., Xu, C.-G., Shao, J.L., Hu, Z.-K.: Phys. Rev. Lett. **117**, 023001 (2016)
21. Möller, C.: The Theory of Relativity. Oxford Univ. Press, London (1952)
22. Lense, J., Thirring, H.: Z. Phys. **19**, 156 (1918); (English translation: B. Mashhoon, Hehl, F.W., Theiss, D.S.: Gen. Rel. Grav. **16**, 711(1984))
23. Weinberg, S.W.: Gravitation and Cosmology. Wiley, New York (1972)
24. Will, C.M.: Theory and Experiment in Gravitational Physics. Cambridge University Press (1993); Will, C.M.: The Confrontation Between General Relativity and Experiment: a 1998 Update, gr-qc/9811036; Camacho, A.: Mod. Phys. Lett. A **16**, 8396 (2001); Barros, A., Romero, C.: Mod. Phys. Lett. A **31**, 2117 (2003)
25. Werner, S.A., Kaiser, H.: Quantum Mechanics in Curved Space-Time. In: Audretsch, J., DeSabbata, V. (eds.) Plenum Press, New York (1990)
26. Papini, G.: Phys. Lett. **23**, 418 (1966); Phys. Lett. **24A**, 32 (1967)
27. Fadeev, P., Wang, T., Band, Y.B., Budker, D., Graham, P.W., Sushkov, A.O., Jackson Kimball, D.F.: Phys. Rev. D **103**, 044056 (2021)
28. Cai, Y.Q., Papini, G.: Mod. Phys. Lett. A 4, 1143 (1989); Gen. Rel. Grav. **22**, 259 (1990)
29. Cai, Y.Q., Papini, G.: Class. Quantum Grav. **7**, 269 (1990)

30. Cai, Y.Q., Papini, G.: Modern Problems of Theoretical Physics, p, 131. In: Pronin, P.I., Obukhov, Y.N. (eds.) World Scientific, Singapore (1991)
31. Szekeres, P.: Nature 228, 1183 (1970). J. Math. Phys. **13**, 286 (1972)
32. Khan, K., Penrose, R.: Nature **229**, 185 (1971)
33. Yurtsever, U.: Phys. Rev. D **38**, 1706 (1988); **38**, 1731 (1988); **40**, 329 (1989)
34. Griffiths, J.B.: Colliding Plane Waves in General Relativity. Oxford University Press, Oxford (1991)
35. Grasso, M., Korzyński, M., Serbenta, J.: arXiv:1811.10284v3 [gr-qc] 7Mar 2019
36. Schneider, P., Ehlers, J., Falco, E.E.: Gravitational Lenses. Springer-Verlag, New York (1992)
37. Deguchi, S., Watson, W.D.: Phys. Rev. D **34**, 1708 (1986)
38. Takahashi, R., Nakamura, T.: ApJ **595**, 1039 (2003)
39. Papini, G.: Relativity in Rotating Frames, p. 335. In: Rizzi, G., Ruggiero, M.L. (eds.) Kluwer Science Publishers, Dordrecht (2004)

Neutrinos in Gravitational Fields 4

Abstract

The chapter is concerned with neutrino optics, helicity, and flavour oscillations induced by gravitational fields. It also deals with neutrino lensing, the spin-gravity coupling of neutrinos with primordial gravitational waves and with the phenomenon of pulsar kick.

4.1 Introduction

The behaviour of spin-1/2 particles in the presence of gravitational fields $g_{\mu\nu}$ is determined by the covariant Dirac equation

$$[i\gamma^{\mu}(x)\mathcal{D}_{\mu} - m]\Psi(x) = 0\,. \tag{4.1}$$

Neutrino beams produced in weak interactions may be considered as a superposition of different mass eigenstates. As a beam propagates, different components of the beam evolve differently so that the probability of finding different eigenstates in the beam varies with time. The consequences of this can be explored in a number of cases.

4.1.1 Neutrino Helicity Oscillations

Consider a beam of high-energy neutrinos. If the neutrino source rotates, the effective Hamiltonian for the mass eigenstates can be written as

$$H_e = (p^2 + m^2)^{\frac{1}{2}} + \Gamma_0 \approx p + \frac{m^2}{2E} - \frac{1}{2}\boldsymbol{\omega}\cdot\boldsymbol{\sigma}\,.$$

G. Lambiase and G. Papini, *The Interaction of Spin with Gravity in Particle Physics*, Lecture Notes in Physics 993, https://doi.org/10.1007/978-3-030-84771-5_4

Consider, for simplicity, a one generation of neutrinos that can be written as a superposition of ν_L and ν_R in the form

$$|\nu(t)>= a_L(t)|\nu_L> +b_R(t)|\nu_R> . \tag{4.2}$$

It is well known that the standard model contemplates only the existence of ν_L, while ν_R is considered sterile and therefore unobservable. Strictly speaking one should consider the helicity states $\nu_\pm$(that are mass eigenstates) in (4.2), however, at high energies

$$\nu_L \simeq \nu_- , \quad \nu_R \simeq \nu_+ .$$

Assuming that $m_1 \neq m_2$, taking $p_1 \sim p_2$ along the x_3-axis and substituting (4.2) into the Schroedinger equation that corresponds to H_e, one obtains

$$i\frac{\partial}{\partial t}\begin{pmatrix} a_L \\ b_R \end{pmatrix} = \begin{pmatrix} p + \frac{m_1^2}{2E} & -\frac{\omega_1}{2} - i\frac{\omega_2}{2} \\ -\frac{\omega_1}{2} + i\frac{\omega_2}{2} & p + \frac{m_2^2}{2} \end{pmatrix}\begin{pmatrix} a_L \\ b_R \end{pmatrix} \equiv M_{12}\begin{pmatrix} a_L \\ b_R \end{pmatrix}. \tag{4.3}$$

The eigenvalues of M_{12} are

$$k_\mp = p + \frac{m_1^2 + m_2^2}{4E} \mp \left[\left(\frac{\Delta m^2}{2E}\right)^2 + \omega_\perp^2\right]^{\frac{1}{2}}, \tag{4.4}$$

where

$$\Delta m^2 \equiv m_1^2 - m_2^2$$

and

$$\omega_\perp^2 \equiv \omega_1^2 + \omega_2^2.$$

The eigenvectors are

$$\begin{aligned} |\nu_1\rangle &= b_1\left[\eta_1|\nu_L\rangle + |\nu_R\rangle\right], \\ |\nu_2\rangle &= b_2\left[\eta_2|\nu_L\rangle + |\nu_R\rangle\right], \end{aligned} \tag{4.5}$$

where

$$\begin{aligned} \eta_1 &= \frac{\omega_1 + i\omega_2}{\Omega + \frac{\Delta m^2}{2E}}, \\ \eta_2 &= \frac{\omega_1 + i\omega_2}{-\Omega + \frac{\Delta m^2}{2E}}, \end{aligned} \tag{4.6}$$

$$\begin{aligned} |b_1|^2 &= \frac{1}{1+|\eta_1|^2}, \\ |b_2|^2 &= \frac{1}{1+|\eta_2|^2}, \end{aligned} \tag{4.7}$$

and

$$\Omega = \left[\left(\frac{\Delta m^2}{2E} \right)^2 + \omega_\perp^2 \right]^{\frac{1}{2}}. \tag{4.8}$$

One therefore finds

$$|\nu(t)> = \frac{b_1}{\eta_1 - \eta_2} exp\left[-i \left(p + \frac{m_1^2 + m_2^2}{4E} \right) t \right] \times$$
$$\left[\left(e^{i\frac{\Omega}{2}t}\eta_1 - e^{-i\frac{\Omega}{2}t}\eta_2 \right) \nu_L + 2i \sin(\frac{\Omega t}{2})\nu_R \right], \tag{4.9}$$

where the initial condition is $\nu(0) = \nu_L$. The transition probability is

$$P_{\nu_L \to \nu_R} = | < \nu_R|\nu(t) > |^2 = \frac{\omega_\perp^2}{2\Omega^2}[1 - \cos(\Omega t)]. \tag{4.10}$$

If the neutrinos have mass, then the magnitude of the transition probability becomes appreciable if

$$\omega_\perp \geq \frac{\Delta m^2}{2E}. \tag{4.11}$$

Unlike the flavour oscillations generated by the MSW mechanism [1,2] that require $\Delta m^2 \neq 0$, helicity oscillations can also occur when $m_1 = m_2$. They are interesting because ν_R's, if they exist, do not interact with matter and would therefore provide an energy dissipation mechanism with possible astrophysical implications. The conversion rate $\nu_L \to \nu_R$ is not large for galaxies and white dwarfs. Assume in fact that $\omega_\perp \gg \Delta m^2/2E$ and that the beam of neutrinos consists of $N_L(0)$ particles at $z = 0$. One immediately derives from (4.10) that the relative numbers of particles at $z = 0$ are [3]

$$N_L(z) = N_L(0)\cos^2\left(\frac{\omega_\perp z}{2c}\right), N_R(z) = N_L(0)\sin^2\left(\frac{\omega_\perp z}{2c}\right). \tag{4.12}$$

One obtains from (4.12) $N_R \sim 10^{-6} N_L(0)$ for galaxies of typical size L such that $\omega_\perp L \sim 200$ km/s. Similarly, for white dwarfs $\omega_\perp \sim 1.0s^{-1}$ and one finds $N_R \sim 10^{-4} N_L(0)$. In the case of Sun, $\omega_\perp \sim 7.3 \times 10^{-5} - 2.4 \times 10^{-6} s^{-1}$ and the conversion rate peaks at distances $L \sim 10^{15} - 4 \times 10^{16}$ cm, well in excess of the average Sun–Earth distance. Helicity oscillations could not therefore explain the solar neutrino puzzle without additional assumptions about the Sun's structure [4]. For neutron stars, however, the dynamics of the star could be affected by this cooling mechanism. In fact, neutrinos diffuse out of a canonical neutron star in a time 1 to 10s, during which they travel a distance 3×10^9 cm between collisions. At distances $L \sim 5 \times 10^6$ cm (the star's radius), the conversion rate is $N_R \sim 0.5 N_0$. Even higher cooling rates may occur at higher rotational speeds and prevent the formation of a pulsar. These results do not require the existence of a magnetic moment for the neutrino (which would also require some mass). Its effect could be taken into account by adding the term $\boldsymbol{\mu} \cdot \boldsymbol{B}$ to H_e. In all instances considered, however, magnetic spin-flip rates of magnitude comparable to those discussed require neutrino magnetic moments vastly in excess of the value $10^{-19}\mu_0\left(\frac{m_\nu}{1eV}\right)$ predicted by the standard $SU(2) \times U(1)$ electro-weak theory [5].

Neutrino spin oscillations in a Schwarzschild metric have been calculated explicitly in [6], with particular attention to the process on different circular orbits.

4.1.2 Helicity Oscillations in a Medium

The behaviour of neutrinos in a medium is modified by a potential V. Subtracting from the diagonal terms of the Hamiltonian a common factor which contributes only to the overall phase and is therefore irrelevant, the effective Hamiltonian becomes

$$H = p + \frac{m^2}{2E} - V - \frac{1}{2}\boldsymbol{\omega} \cdot \boldsymbol{\sigma}. \tag{4.13}$$

For simplicity, consider again one-generation neutrinos and assume that V is constant. Applying the diagonalization procedure of the last section to the new Hamiltonian leads to the transition probability

$$P_{\nu_L \to \nu_R} = \frac{\omega_\perp^2}{2\Omega'^2}[1 - \cos(\Omega' t)], \tag{4.14}$$

where $\Omega' = [(V + \frac{\Delta m^2}{2E})^2 + \omega_\perp^2]^{\frac{1}{2}}$. From (4.14), one finds that spin-flip transitions are strongly suppressed when $V + \frac{\Delta m^2}{2E} > \omega_\perp$ and only the ν_L component is present in the beam. If $\omega_\perp > V + \frac{\Delta m^2}{2E}$, then the ν_L flux has an effective modulation. Resonance occurs at $V = \frac{-\Delta m^2}{2E}$. Consider now the rotating core of a supernova. In this case, V can be relatively large of the order of several electron volts and describes the interaction of neutrinos with the particles of the medium. For right-handed neutrinos V vanishes. Assuming that the star does not radiate more energy as ν_R's than as ν_L's, one finds $L_{\nu_L} \sim L_{\nu_R} \sim 5 \times 10^{53} erg/s$. As the star collapses, spin-rotation coupling acts on both ν_L and ν_R. The ν_L's become trapped and leak towards the exterior ($l \sim 1.5 \times 10^7$ cm), while their interaction with matter is $V \sim 14(\rho/\rho_c)$ eV and increases with the medium's density, which is $\rho_c \sim 4 \times 10^{14} g/\text{cm}^3$ at the core. The ν_R's escape. As ρ increases, the transition $\nu_L \to \nu_R$ is inhibited (off resonance). One also finds $\frac{\Delta m^2}{2E} < \omega_\perp$ when $\Delta m^2 < 10^{-5} eV^2$, $E \sim 10 MeV$, $\omega_\perp \sim 6 \times 10^3 s^{-1}$. It then follows from (4.14) that

$$L_{\nu_L} \sim L_{\nu_R} \sin^2\left(\frac{\omega_\perp l}{2c}\right), \tag{4.15}$$

where $\frac{\omega_\perp l}{2c} \sim \frac{\pi}{2}$. In the time $\frac{l}{c} \sim 5 \times 10^{-4} s$, the energy associated with the $\nu_R \to \nu_L$ conversion is $\sim 2.5 \times 10^{50} erg$ which is just the missing energy required to blow up the mantle of the collapsing star [3].

4.1.3 Neutrino Flavour Oscillations: The Effect of Neutrino's Travel Time

Consider a beam of neutrinos of fixed energy E emitted at point (r_A, t_A) of the (r, t)-plane and assume that the particles are in a weak flavour eigenstate, that is, in a linear superposition of mass eigenstates m_1 and m_2, with $m_1 \neq m_2$. It is argued in the literature [7,8] that if interference is observed at the same space–time point (r_B, t_B), then the lighter component must have left the source at a later time $\Delta t = \frac{r_B - r_A}{v_1} - \frac{r_B - r_A}{v_2}$, where v_1 and v_2 are the velocities of the eigenstates of masses m_1 and m_2, respectively. Because of the difference in travel time Δt, gravity-induced neutrino flavour oscillations will ensue even though gravity couples universally to matter. Ignoring spin contributions, the phase difference of the two mass eigenstates can be calculated in a completely gauge-invariant way. Assume the neutrinos propagate in a gravitational field described by the Schwarzschild metric. When the closed space–time path in the solution of the Klein–Gordon equation is extended to the triangle (r_A, t_A), (r_B, t_B), $(r_A, t_A + \Delta t)$, one obtains

$$
\begin{aligned}
&(i\Delta\Phi_G\phi_0)_{m_1} - (i\Delta\Phi_G\phi_0)_{m_2} = \\
&= \frac{r_g E}{2}[-\frac{v_1 \Delta t}{2} - \frac{1}{v_1}\ln(\frac{r_B}{r_A + v_1 \Delta t}) + (-v_1 + \frac{1}{v_2} + v_2)\ln\frac{r_B}{r_A}] \simeq \\
&\simeq \frac{r_g E}{2}(\frac{1}{v_2} - \frac{1}{v_1} + v_2 - v_1)\ln\frac{r_B}{r_a},
\end{aligned}
\tag{4.16}
$$

where the approximation $v_1 \Delta t \ll r_A$ has been used in deriving the last result. On using the relation $1/v = E/p$ and the approximations $v \sim 1 - \frac{m^2}{2E^2} - \frac{m^4}{8E^4}$, $\frac{1}{v} \sim 1 + \frac{m^2}{2E^2} + \frac{3m^4}{8E^4}$, one arrives at the result

$$
\begin{aligned}
&(i\Delta\Phi_G\phi_0)_{m_1} - (i\Delta\Phi_G\phi_0)_{m_2} \simeq \\
&\simeq \frac{MGc^5}{4\hbar E^3}(m_2^4 - m_1^4)\ln\frac{r_B}{r_A} = \\
&= 1.37 \times 10^{-19}\frac{M}{M_\odot}\frac{\Delta m^4}{eV^4}\left(\frac{MeV}{E}\right)^3 \ln\frac{r_B}{r_A}.
\end{aligned}
\tag{4.17}
$$

The effect therefore exists, but is small in typical astrophysical applications.

Torsion-induced neutrino oscillations have been considered by de Sabbata and collaborators [9,10]. Neutrino spin and flavour oscillations have also been studied in alternative theories of gravity [11].

4.2 Neutrino Optics

When neutrinos propagate near a rotating gravitational source, lensing may occur. The deflection angle δ is defined as

$$\tan\delta = \frac{\sqrt{-g_{ij}p_{\perp}^{i}p_{\perp}^{j}}}{p_{\parallel}}, \tag{4.18}$$

where $\boldsymbol{p}_{\perp}$ and $p_{\parallel}$ are the orthogonal and parallel components of the momentum with respect to the initial direction of neutrinos. In the weak field approximation $\tan\varphi \simeq \varphi$ and (4.18) reduces to

$$\delta \simeq \frac{|\boldsymbol{p}_{\perp}|}{k_{\parallel}}, \tag{4.19}$$

where $k_{\parallel} = p_{\parallel}$ is the unperturbed momentum and $|\boldsymbol{p}_{\perp}| = \sqrt{-\eta_{ij}p_{\perp}^{i}p_{\perp}^{j}}$, with $p_{\perp}^{i} \sim O(\gamma_{\mu\nu})$.

The vector $\boldsymbol{p}_{\perp}$ is derived by taking the T_{0i} components of the neutrinos energy–momentum tensor

$$p_i \equiv \int d^3x\sqrt{-g}\; T^{0}_{\ i}(x), \tag{4.20}$$

where [12]

$$T_{\mu\nu} = \frac{i}{4}\left[\bar{\Psi}\gamma_{\mu}\nabla_{\nu}\Psi + \bar{\Psi}\gamma_{\nu}\nabla_{\mu}\Psi - (\nabla_{\mu}\bar{\Psi})\gamma_{\nu}\Psi - (\nabla_{\nu}\bar{\Psi})\gamma_{\mu}\Psi\right]. \tag{4.21}$$

Here $\bar{\Psi} = \Psi^{\dagger}\gamma^{\hat{0}}$, and Ψ is given by (1.73), which to first order in $\gamma_{\mu\nu}$ can be cast in the form

$$\Psi = \Psi_0 + \delta\Psi, \tag{4.22}$$

with

$$\delta\Psi = \frac{1}{2m}\left[-i2m\Phi_G - i(\gamma^{\hat{\mu}}k_{\mu} + m)\Phi_s + \gamma^{\hat{\alpha}}(h^{\mu}_{\hat{\alpha}}k_{\mu} + \Phi_{G,\alpha})\right]\Psi_0, \tag{4.23}$$

and

$$e^{\mu}_{\hat{\alpha}} \simeq \delta^{\mu}_{\alpha} + h^{\mu}_{\hat{\alpha}}. \tag{4.24}$$

The gravitational contributions are contained exclusively in the terms $h^{\mu}_{\hat{\alpha}}$ and $k_{\mu} = (E, \boldsymbol{k})$ is the unperturbed neutrino momentum. The components $\boldsymbol{p}_{\perp}$ in (4.19) are evaluated by means of $T^{0}_{\ i}$, according to (4.20), where the index i fixes the directions orthogonal to the propagation of the neutrinos.

The expression for $T^0_{\,i}$ can be simplified considerably by assuming that only the component of $\boldsymbol{k}$ in the direction of propagation of the neutrinos does not vanish. Then

$$T^0_{\,i} = \frac{i}{4}\Big[i\bar{\Psi}_0\{\gamma^{\hat{0}}, \Gamma_i\}\Psi_0 + (\bar{\Psi}_0\gamma^{\hat{0}}\partial_i\delta\Psi - \partial_i\delta\bar{\Psi}\gamma^{\hat{0}}\Psi_0)$$
$$-h^i_{\hat{\mu}}(\bar{\Psi}_0\gamma^{\hat{\mu}}\partial_0\Psi_0 - \partial_0\bar{\Psi}_0\gamma^{\hat{\mu}}\Psi_0) - i\bar{\Psi}_0\{\gamma^{\hat{i}}, \Gamma_0\}\Psi_0 - (\delta\bar{\Psi}\gamma^{\hat{i}}\partial_0\Psi_0 - \partial_0\bar{\Psi}_0\gamma^{\hat{i}}\delta\Psi)$$
$$-(\bar{\Psi}_0\gamma^{\hat{i}}\partial_0\delta\Psi - \partial_0\delta\bar{\Psi}\gamma^{\hat{i}}\Psi_0) + \gamma_{i\mu}(\bar{\Psi}_0\gamma^{\hat{\mu}}\partial_0\Psi_0 - \partial_0\bar{\Psi}_0\gamma^{\hat{\mu}}\Psi_0)\Big]\,. \tag{4.25}$$

In the geometrical optics approximation, for which the uncertainties on measurements largely exceed the limits imposed by the Heisenberg inequality, the neutrinos can be considered as wave packets with a negligible spread both in coordinate and momentum space. One can thus construct, from the plane-wave solutions of the Dirac equation in flat space–time $\nu_0 e^{-ik_\alpha x^\alpha}$, a (e.g. Gaussian) wave packet $\Psi_0(x)$ whose spatial behaviour is

$$\Psi_0 \sim \alpha^{-1/2} e^{i\langle p\rangle x - (x-vt)^2/2\alpha^2},$$

where α is related to the spread of momentum components ($\Delta p \sim 1/\alpha$). In curved space–time, neutrinos are described by a $\Psi(x)$, which matches Ψ_0 as neutrinos are localized far from the gravitational sources. $\Psi(x)$ still follows a Gaussian profile centred at $\bar{x} = vt$. In WFA, one may assume with good approximation that gravity does not substantially alter the spatial coherence of the wave packet at any generic point $\bar{x}$ where the neutrinos are localized. Within this framework, the products of wave functions in (4.21), or (4.25), give rise to terms proportional to

$$|\Psi(x)|^2 \sim \alpha^{-1} e^{-(x-\bar{x})^2/\alpha^2},$$

which in the limit $\alpha \to 0$ behaves like

$$|\Psi(x)|^2 \sim \delta(x - \bar{x}).$$

As a consequence, Eq. (4.20) implies $p_i \approx \sqrt{-g}\, T^0_i$. Using Eqs. (4.25) and (4.23), one gets

$$(\boldsymbol{p}_\perp)_i = \Phi_{G,i} + \frac{1}{2}h^\mu_{\hat{i}}k_\mu + \frac{E}{2m}\bar{\nu}_0\Gamma_i\nu_0 + \frac{1}{4}\bar{\nu}_0\{\gamma^{\hat{i}}, \Gamma_0\}\nu_0 - \frac{1}{4}\bar{\nu}_0\{\gamma^{\hat{0}}, \Gamma_i\}\nu_0 -$$
$$-\frac{1}{4m}\Phi_{G,\mu 0}\bar{\nu}_0\sigma^{\hat{i}\hat{\mu}}\nu_0 - \frac{1}{2}\gamma_{i\mu}k^\mu\,, \tag{4.26}$$

where the simplified notation $\Psi_0 \equiv \nu_0$ has been used for both left/right-handed neutrinos

$$\Psi_0(x) = \nu_{0L,R}e^{-ik_\alpha x^\alpha} = \sqrt{\frac{E+m}{2E}}\begin{pmatrix}\nu_{L,R}\\ \frac{\sigma\cdot k}{E+m}\nu_{L,R}\end{pmatrix}e^{-ik_\alpha x^\alpha}\,. \tag{4.27}$$

In (4.27) $\boldsymbol{\sigma} = (\sigma^1, \sigma^2, \sigma^3)$ are the Pauli matrices and

$$\nu_L = \begin{pmatrix} 0 \\ 1 \end{pmatrix}, \quad \nu_R = \begin{pmatrix} 1 \\ 0 \end{pmatrix}.$$

Notice that

$$\bar{\nu}_{0\,L,R}(k) \equiv \nu^{\dagger}_{0\,L,R}(k)\gamma^{\hat{0}}, \quad \nu^{\dagger}_{0\,L,R}(k)\nu_{0\,L,R}(k) = 1.$$

Using the results of Sect. (1.3.5) for the Lense–Thirring metric [13] one obtains the spin connection (1.61). In what follows, neutrinos can propagate along the z-axis, parallel to the angular velocity of the gravitational source, and along the x-axis, orthogonal to the angular velocity of the gravitational source.

Propagation Along the z-Axis

In Eq. (4.26) and (4.27), $\nu_L = \begin{pmatrix} 0 \\ 1 \end{pmatrix}$, that is, ν_L is an eigenstate of σ^3. Neutrinos start from $z = -\infty$ with impact parameter b. Without loss of generality, one can choose neutrinos propagating along $x = b$, $y = 0$.

From (4.26), one gets for $i = 1, 2$, to order $O(m^2/E^2)$,

$$(p_\perp)_1 = \frac{2GMk^3}{b}\left(1 + \frac{m^2}{2E^2}\right)\left(1 + \frac{z}{r}\right), \tag{4.28}$$
$$(p_\perp)_2 = \frac{4GMR^2\omega bk^3}{5r^3}\left(1 + \frac{m^2}{2E^2}\right).$$

Returning to (4.19), taking $k_\parallel = k^3$ and $(p_\perp)_{1,2}$ as in (4.28), one finds

$$\delta = \frac{2GM}{b}\left(1 + \frac{m^2}{2E^2}\right)\sqrt{\left(1 + \frac{z}{r}\right)^2 + \left(\frac{2R^2b^2\omega}{r^3}\right)^2}, \tag{4.29}$$

which is the deflection predicted by General Relativity for photons up to corrections due to the neutrino mass. In the limit $z \to \infty$ Eq. (4.29) reduces to

$$\delta = \frac{4GM}{b}\left(1 + \frac{m^2}{2E^2}\right). \tag{4.30}$$

Propagation Along the x-Axis

To analyse the propagation along the x-axis, one considers Eqs. (4.26), and (4.27) with $\nu_L = \frac{1}{\sqrt{2}}\begin{pmatrix} 1 \\ -1 \end{pmatrix}$ (i.e. ν_L is the eigenstate of σ^1). Neutrinos start from $x = -\infty$

with impact parameter b. For the sake of simplicity, consider neutrinos propagating in the equatorial plane $z = 0$, $y = b$.

Equation (4.26) for $i = 2, 3$ gives, to order $O(m^2/E^2)$,

$$(p_\perp)_2 = \frac{2GMk^1}{b}\left(1 - \frac{2R^2\omega}{5b}\right)\left(1 + \frac{m^2}{2E^2}\right)\left(1 + \frac{x}{r}\right), \tag{4.31}$$
$$(p_\perp)_3 = 0 .$$

From (4.19), and for $k_\parallel = k^1$ and $(p_\perp)_{2,3}$ given by (4.31), one gets

$$\delta = \frac{2GM}{b}\left(1 - \frac{2R^2\omega}{5b}\right)\left(1 + \frac{m^2}{2E^2}\right)\left(1 + \frac{x}{r}\right), \tag{4.32}$$

which contains, unlike the case of propagation along the z-axis, the contribution of the angular momentum of the gravitational source. In the limit of vanishing mass, the result coincides with the lensing angle for photons, as it follows from geometrical optics. It also follows that in the limit $z \to \infty$

$$\delta = \frac{4GM}{b}\left(1 - \frac{2R^2\omega}{5b}\right)\left(1 + \frac{m^2}{2E^2}\right). \tag{4.33}$$

4.3 Helicity Transitions Induced by Gravitational Fields

To study the helicity flip of neutrinos, it is convenient to write the neutrino wave function in the form

$$|\psi(\lambda)\rangle = \alpha(\lambda)|\nu_R\rangle + \beta(\lambda)|\nu_L\rangle , \tag{4.34}$$

where $|\alpha|^2 + |\beta|^2 = 1$ and λ is an affine parameter along the world line. At $\lambda + d\lambda$, (4.34) becomes

$$|\psi(\lambda + d\lambda)\rangle = \left(\alpha + \frac{d\alpha}{d\lambda}\right)|\nu_R\rangle + \left(\beta + \frac{d\beta}{d\lambda}\right)|\nu_L\rangle . \tag{4.35}$$

To determine $d\alpha/d\lambda$ and $d\beta/d\lambda$, one writes Eq. (1.72) as [14]

$$|\psi(\lambda)\rangle = \hat{T}(\lambda)|\psi_0\rangle , \tag{4.36}$$

where

$$\hat{T} = -\frac{1}{2m}\left(-i\gamma^\mu(x)\mathcal{D}_\mu - m\right)e^{-i\Phi_T} , \tag{4.37}$$

and $|\psi_0\rangle$ is the corresponding solution in Minkowski space–time. Evaluating (4.36) at $\lambda + d\lambda$, one gets

$$\begin{aligned} |\psi(\lambda + d\lambda)\rangle &\simeq \hat{T}(\lambda)|\psi_0\rangle + \dot{x}^\mu\partial_\nu\hat{T}|\psi_0(\lambda)\rangle d\lambda \\ &\simeq |\psi(\lambda)\rangle + \dot{x}^\mu\partial_\nu\hat{T}|\psi(\lambda)\rangle d\lambda . \end{aligned} \tag{4.38}$$

To first order $|\psi(\lambda)\rangle \simeq |\psi_0\rangle$ because $\partial_\mu \hat{T}$ is already of the first order in $\gamma_{\mu\nu}$. One can then write

$$\frac{|\psi(\lambda + d\lambda)\rangle - |\psi(\lambda)\rangle}{d\lambda} \simeq \dot{x}^\mu \partial_\mu \hat{T} |\psi(\lambda)\rangle \,. \tag{4.39}$$

It follows from (4.34) and (4.35) that

$$\begin{aligned} \frac{d\alpha}{d\lambda} &= \alpha\, \langle \nu_R | \dot{x}^\mu \partial_\mu \hat{T} | \nu_R \rangle + \beta\, \langle \nu_R | \dot{x}^\mu \partial_\mu \hat{T} | \nu_L \rangle \,, \\ \frac{d\beta}{d\lambda} &= \alpha\, \langle \nu_L | \dot{x}^\mu \partial_\mu \hat{T} | \nu_R \rangle + \beta\, \langle \nu_L | \dot{x}^\mu \partial_\mu \hat{T} | \nu_L \rangle \,. \end{aligned} \tag{4.40}$$

If the neutrinos are created in a quasi left-handed state, i.e. $|\alpha(0)|^2 \ll |\beta(0)|^2$, one obtains

$$\frac{d\alpha}{d\lambda} \simeq \langle \nu_R | \dot{x}^\mu \partial_\mu \hat{T} | \nu_L \rangle \,. \tag{4.41}$$

The integration yields the transition probability

$$P_{\nu_L \to \nu_R} = |\alpha(\lambda)|^2 = | \int_{\lambda_0}^{\lambda} \langle \nu_R | \dot{x}^\mu \partial_\mu \hat{T} | \nu_L \rangle d\lambda |^2 \,. \tag{4.42}$$

Choosing $\lambda = t$, so that $\dot{x}^\mu = k^\mu / E$ and noting that

$$\partial_\mu \hat{T} = \frac{1}{2m} \left(-i2m\Phi_{G,\lambda} - i(\gamma^{\hat{\mu}} k_\mu + m)\hat{\Phi}_{s,\lambda} + h^\mu_{\hat{\alpha},\lambda} k_\mu \gamma^{\hat{\alpha}} + \gamma^{\hat{\mu}} \Phi_{G,\,\mu\lambda} \right) , \tag{4.43}$$

$$\hat{\Phi}_{s,\lambda} = \Gamma_\lambda \,, \quad \hat{\Phi}^\dagger_{s,\lambda} = \gamma^0 \Gamma_\lambda \gamma^0 \,, \quad \Phi_{G,\mu\lambda} = k_\alpha \Gamma^\alpha_{\mu\lambda} \,, \quad \nu_0^\dagger (\gamma^\mu k_\mu + m) = 2E \nu_0^\dagger \gamma^{\hat{0}} \,,$$

one finds

$$\langle \nu_R | \dot{x}^\mu \partial_\mu \hat{T} | \nu_L \rangle = -i \,\frac{k^\lambda}{m} \bar{\nu}_R \Gamma_\lambda \nu_L + \frac{k^\lambda k_\mu}{2mE} (h^\mu_{\hat{\alpha},\,\lambda} + \Gamma^\mu_{\alpha\lambda}) \nu_R^\dagger \gamma^{\hat{\alpha}} \nu_L \,. \tag{4.44}$$

The probability amplitude (4.44) can now be computed for neutrinos propagating along the z- and x-directions.

Propagation Along z

Setting $k^0 = E$, $k^3 \simeq E(1 - m^2/2E^2)$ and, without loss of generality, $x = b$ and $y = 0$, Eq. (4.44) gives

$$\begin{aligned} \langle \nu_R | \dot{x}^\mu \partial_\mu \hat{T} | \nu_L \rangle = {} & -i \,\frac{E}{m} \left[\bar{\nu}_R \Gamma_0 \nu_L + \left(1 - \frac{m^2}{2E^2} \right) \bar{\nu}_R \Gamma_3 \nu_L \right] + \\ & + \frac{E}{2m} \left[\Gamma^0_{\alpha 0} + \Gamma^0_{\alpha 3} - \Gamma^3_{\alpha 0} - \Gamma^3_{\alpha 3} + h^0_{\hat{\alpha},\,3} - h^3_{\hat{\alpha},\,3} \right] \nu_R^\dagger \gamma^{\hat{\alpha}} \nu_L \end{aligned} \tag{4.45}$$

$$-\frac{m}{4E}\left[-\Gamma^3_{\alpha 0}+\Gamma^0_{\alpha 3}-2\Gamma^3_{\alpha 3}-h^0_{\hat\alpha,3}-2h^3_{\hat\alpha,3}\right]\nu_R^\dagger\gamma^{\hat\alpha}\nu_L\,.$$

Using (1.61) and the equations

$$\bar\nu_R\Gamma_0\nu_L = i\frac{k^3}{E}\frac{GMx}{2r^3}-\frac{3GMR^2\omega xz}{5r^5}\,, \tag{4.46}$$

$$\begin{aligned}
\bar\nu_R\Gamma_3\nu_L &= -\frac{k^3}{E}\frac{3GMR^2\omega xz}{5r^5}+i\frac{GMx}{2r^3}\,,\\
\nu_R^\dagger\gamma^{\hat 1}\nu_L &= -\frac{k^3}{E}\,,\quad \nu_R^\dagger\gamma^{\hat 2}\nu_L = i\frac{k^3}{E}\,,\\
\nu_R^\dagger\gamma^{\hat 0}\nu_L &= \nu_R^\dagger\gamma^{\hat 3}\nu_L = 0\,,
\end{aligned} \tag{4.47}$$

one gets

$$\frac{d\alpha}{d\lambda}\simeq\langle\nu_R|\dot x^\mu\partial_\mu\hat T|\nu_L\rangle=\frac{m}{E}\frac{GMx}{2r^3}\left(1+i\,\frac{6R^2\omega z}{5r^2}\right). \tag{4.48}$$

Contributions to order $O((m/E)^0)$ vanish, so that the probability amplitude for the $\nu_L\to\nu_R$ transition is of order $O(m/E)$.

Integrating from $-\infty$ to z one finds, to order $O(m/E)$,

$$\alpha\simeq\frac{m}{E}\frac{GM}{2b}\left[1+\frac{z}{r}-i\,\frac{2\omega R^2b^2}{5r^3}\right], \tag{4.49}$$

hence

$$P_{\nu_L\to\nu_R}(-\infty,z)=|\alpha|^2\simeq\left(\frac{m}{E}\right)^2\left(\frac{GM}{2b}\right)^2\left[\left(1+\frac{z}{r}\right)^2+\left(\frac{2\omega b^2R^2}{5r^3}\right)^2\right]. \tag{4.50}$$

As expected, the transition probability is of order $O(m^2/E^2)$. The first term comes from the mass of the gravitational field source and the second from its angular velocity. As a consequence, the transition from a neutrino left to neutrino right can be also induced by a non-rotating gravitational source.

The term due to the angular momentum of the source is odd and decreases for $r\to\infty$, thus the contribution from 0 to r is cancelled out by the contribution from $-\infty$ to 0. On the contrary, for neutrinos propagating from 0 to $+\infty$, one gets

$$P_{\nu_L\to\nu_R}(0,+\infty)=|\alpha|^2\simeq\left(\frac{m}{E}\right)^2\left(\frac{GM}{2b}\right)^2\left[1+\left(\frac{2\omega R^2}{5b}\right)^2\right]. \tag{4.51}$$

In this case too, the term depending on the mass of the source is dominant with respect to that depending on its angular velocity.

Propagation Along x

In order to evaluate the transition probability for $\nu_L \to \nu_R$ when neutrinos propagate along the x-direction, it is convenient to choose $k^\mu = (E, k^1, 0, 0)$, with $k^1 \simeq E(1 - m^2/2E^2)$, and $z = 0$, $y = b$. Equation (4.44) then becomes

$$\langle \nu_R | \dot{x}^\mu \partial_\nu \hat{T} | \nu_L \rangle = -i\frac{E}{m}\left[\bar{\nu}_R \Gamma_0 \nu_L + \left(1 - \frac{m^2}{2E^2}\right)\bar{\nu}_R \Gamma_1 \nu_L\right] + \tag{4.52}$$

$$+\frac{E}{2m}\left[\Gamma^0_{\alpha 0} + \Gamma^0_{\alpha 1} - \Gamma^1_{\alpha 0} - \Gamma^1_{\alpha 1} + h^0_{\hat{\alpha},1} - h^1_{\hat{\alpha},1}\right]\nu_R^\dagger \gamma^{\hat{\alpha}} \nu_L$$

$$-\frac{m}{4E}\left[-\Gamma^1_{\alpha 0} + \Gamma^0_{\alpha 1} - 2\Gamma^1_{\alpha 1} + h^0_{\hat{\alpha},1} - 2h^1_{\hat{\alpha},1}\right]\nu_R^\dagger \gamma^{\hat{\alpha}} \nu_L \,.$$

Using (1.61) one obtains

$$\begin{aligned}
\bar{\nu}_R \Gamma_0 \nu_L &= \frac{k^1}{E}\frac{GMy}{2r^3} + \frac{GMR^2\omega}{5r^3}\,, \\
\bar{\nu}_R \Gamma_3 \nu_L &= \frac{k^1}{E}\frac{GMR^2\omega}{5r^3} - \frac{GMy}{2r^3}\,, \\
\nu_R^\dagger \gamma^{\hat{2}} \nu_L &= -i\frac{k^1}{E}\,, \quad \nu_R^\dagger \gamma^{\hat{3}} \nu_L = -\frac{k^1}{E}\,, \\
\nu_R^\dagger \gamma^{\hat{0}} \nu_L &= \nu_R^\dagger \gamma^{\hat{1}} \nu_L = 0\,,
\end{aligned} \tag{4.53}$$

and finally

$$\frac{d\alpha}{dt} = \langle \nu_R | \dot{x}^\mu \partial_\mu \hat{T} | \nu_L \rangle = i\frac{m}{E}\frac{GMb}{2r^3}\left(1 - \frac{2R^2\omega}{5b}\right). \tag{4.54}$$

The contributions to order $O((m/E)^0)$ again vanish and the probability amplitude for the $\nu_L \to \nu_R$ transition is of order $O(m/E)$ and, as Eq. (4.48) shows, the transition $\nu_L \to \nu_R$ can also be induced by non-rotating gravitational sources.

Integrating (4.54) from $-\infty$ to x, one gets

$$\alpha \simeq i\frac{m}{E}\frac{GM}{2b}\left(1 - \frac{2\omega R^2}{5b}\right)\left(1 + \frac{x}{r}\right), \tag{4.55}$$

hence the transition probability $P_{\nu_L \to \nu_R}$ to $O(m^2/E^2)$ is

$$P_{\nu_L \to \nu_R}(-\infty, x) \sim \left(\frac{m}{E}\right)^2 \left(\frac{GM}{2b}\right)^2 \left(1 - \frac{2\omega R^2}{5b}\right)^2 \left(1 + \frac{x}{r}\right)^2. \tag{4.56}$$

In this case too the dominant term is related to the mass of the gravitational source. Notice, however, that the angular momentum of the source in (4.56) is even and does not decrease for $r \to \infty$. The contribution from 0 to x must therefore be added to that from $-\infty$ to 0.

When the neutrinos are generated at $x = 0$ and propagate up to $x = +\infty$, one finds

$$P_{\nu_L \to \nu_R}(0, +\infty) \sim \left(\frac{m}{E}\right)^2 \left(\frac{GM}{2b}\right)^2 \left(1 - \frac{2\omega R^2}{5b}\right)^2 , \tag{4.57}$$

which reduces to

$$P_{\nu_L \to \nu_R}(0, +\infty) \sim \left(\frac{m}{E}\right)^2 \left(\frac{GM}{2b}\right)^2 \left(1 - \frac{4\omega R^2}{5b}\right) , \tag{4.58}$$

in the approximation $\omega R^2/b \ll 1$.

4.4 Neutrino Flavour Oscillations

It is convenient to rewrite the first-order solution of the covariant Dirac equation in the form

$$\Psi(x) = f(x) e^{-i\Phi_T - ik_\alpha x^\alpha} \Psi_0 , \tag{4.59}$$

where

$$f(x) = \frac{1}{2m}\left[e^{\mu}_{\hat{\alpha}} \gamma^{\hat{\alpha}} (k_\mu + \Phi_{G,\mu}) + m\right] , \tag{4.60}$$

and Ψ_0 is given by the phase-independent part of $\Psi(x)$ in the case of mass eigenstate neutrinos. The relationship between flavour eigenstates (Greek indices) and mass eigenstates (Latin indices) is given by the standard expression

$$|\nu_\alpha(x)\rangle = \sum_j U_{\alpha j}(\theta) |\nu_j(x)\rangle , \tag{4.61}$$

into which (4.59) must now be substituted. The result is

$$|\nu_\alpha(x(\lambda))\rangle = \sum_j U_{\alpha j}(\theta) f_j e^{-i\tilde{\Phi}^j_T(x) - ik^j_\alpha x^\alpha} |\nu^j\rangle \equiv \sum_j U_{\alpha j} \hat{O}(x(\lambda)) |\nu^j\rangle , \tag{4.62}$$

where $\Phi_T = \Phi_s + \Phi_G$,

$$\hat{O}(\lambda) = \frac{1}{2m}\left[e^{\mu}_{\hat{\alpha}} \gamma^{\hat{\alpha}} (k_\mu + \Phi_{G,\mu}) + m\right] e^{-i(\Phi_T + k\cdot x)} , \tag{4.63}$$

and U is the mixing matrix

$$U = \begin{pmatrix} \cos\theta & \sin\theta \\ -\sin\theta & \cos\theta \end{pmatrix} . \tag{4.64}$$

It is worth noticing that a mass eigenstate $|\nu_i\rangle$ is described by the vector state

$$|\nu_i\rangle = |\zeta_i\rangle \otimes |\varsigma_i\rangle\,, \tag{4.65}$$

where $|\varsigma_i\rangle$ is defined in the mass eigenstate Hilbert space $\langle\varsigma_i|\varsigma_j\rangle = \delta_{ij}$, whereas $|\zeta_i\rangle$ contains all information on the nature of a neutrino, such as neutrino helicity or the dependence on the momentum p^μ and coordinates x^μ (in the coordinate representation, one has $\Psi_{0i}(x) = \langle x|\zeta_i\rangle$, which is given by (4.27)).

Restricting α to e for simplicity, one can define the matrix

$$\chi = \begin{pmatrix} \langle\nu_e|\nu_e(x)\rangle \\ \langle\nu_\mu|\nu_e(x)\rangle \end{pmatrix} = \begin{pmatrix} \cos^2\theta\,\hat{O}_{11} + \sin\theta\cos\theta\,\hat{O}_{12} + \sin\theta\cos\theta\,\hat{O}_{21} + \sin^2\theta\,\hat{O}_{22} \\ -\sin\theta\cos\theta\,\hat{O}_{11} - \sin^2\theta\,\hat{O}_{12} + \cos^2\theta\,\hat{O}_{21} + \sin\theta\cos\theta\,\hat{O}_{22} \end{pmatrix} =$$
$$= U^\dagger \begin{pmatrix} O_{11} & O_{12} \\ O_{21} & O_{22} \end{pmatrix} U\phi\,, \tag{4.66}$$

where $\hat{O}_{ij} = \langle\nu_i|\hat{O}(\lambda)|\nu_j\rangle \equiv \nu_i^\dagger \hat{O}\nu_j = \langle\zeta_i|\hat{O}(\lambda)|\zeta_j\rangle\delta_{ij}, i, j = 1, 2,$ and $\phi = \begin{pmatrix} 1 \\ 0 \end{pmatrix}$.

Differentiating with respect to the parameter λ gives

$$i\frac{d\chi}{d\lambda} = U^\dagger \begin{pmatrix} i\frac{dO_{11}}{d\lambda} & 0 \\ 0 & i\frac{dO_{22}}{d\lambda} \end{pmatrix} U\phi\,. \tag{4.67}$$

Keeping only terms of the first order in $\gamma_{\alpha\beta}$ and noticing that $\Phi_{G,\mu\nu} \approx k_\alpha\Gamma^\alpha_{\mu\nu}$, $\Phi_{G,\mu}e^\mu_{\hat{a},\rho} \approx O(\gamma^2_{\mu\nu})$, and $\Phi_{G,\mu}\Gamma_\rho \approx O(\gamma^2_{\mu\nu})$, the matrix elements $i dO_{ij}/d\lambda$ in (4.67) are of the form

$$i\,\frac{dO_{ij}}{d\lambda} \simeq A^{(i)}\,O_{ij} + \sum_k d^{(i)}_{ik}\,O_{kj} + C^{(i)}_{ij}\,, \tag{4.68}$$

where the index (i) refers to mass eigenstates, and

$$A^{(i)} = \dot{x}^{(i)\rho}\left(k^{(i)}_\rho + \Phi^{(i)}_{G,\rho}\right), \quad d^{(i)}_{ij} = \langle\nu_i|\dot{x}^\rho\Gamma_\rho|\nu_j\rangle = \dot{x}^{(i)\rho}\langle\zeta_i|\Gamma_\rho|\zeta_j\rangle\,\delta_{ij}\,, \tag{4.69}$$
$$C^{(i)}_{ij} \equiv i\frac{\dot{x}^{(i)\rho}}{2m_i}(h^\mu_{\hat{\alpha},\rho}k^{(i)}_\mu + \delta^\mu_{\hat{\alpha}}\Phi^{(i)}_{G,\mu\rho})\langle\zeta_i|\gamma^{\hat{\alpha}}|\zeta_j\rangle\,\delta_{ij}\,e^{-i\Phi^{(i)}_G - ik^{(i)}\cdot x}\,.$$

The equation of evolution is therefore

$$i\frac{d\chi}{d\lambda} = (A_f + d_f)\chi + C\,, \tag{4.70}$$

where

$$A_f = U^\dagger \begin{pmatrix} A^{(1)} & 0 \\ 0 & A^{(2)} \end{pmatrix} U\phi\,, \tag{4.71}$$

and

$$C = U^{\dagger} \begin{pmatrix} C_{11} & 0 \\ 0 & C_{22} \end{pmatrix} U\phi\,. \tag{4.72}$$

As discussed below, C can be written as $C = \bar{C}\chi$, where $\bar{C}$ is defined in Eq. (4.115). Since C is a pure geometrical term, it accounts for an overall phase factor which does not contribute to the neutrino oscillations and can be removed in what follows.

The equation of evolution can be transformed into a more convenient form by taking $\dot{x}^{\rho}$ as the tangent vector to the null world line, so that $\dot{x}^{\rho}\dot{x}_{\rho} = 0$, and $k^{\mu} = (E, k^i) \sim (E, n^i E(1 - m^2/2E^2))$ is the neutrino four momentum (n^i is a unit vector parallel to the neutrino three momentum) so that $k^0 = \dot{x}^0$ and $k^i \approx (1-\varepsilon)\dot{x}^i$, with $\varepsilon \ll 1$. In order to observe oscillations, one must assume that neutrinos propagate along null geodesics; it may otherwise happen that mass eigenstates are observed, in different positions or times, destructing in the process the interference pattern. Thus, the matrix components in (4.71) contain terms of the form

$$\dot{x}^{(i)\rho} k_{\rho}^{(i)} \approx \frac{m_i^2}{2} + \varepsilon E^2\,,$$

which are flavour independent, are diagonal in the matrix of evolution, and do not therefore contribute to the oscillations. The mechanical momentum of the neutrino along its world line can be written in the form

$$P_s = (k_{\rho} + \Phi_{T,\rho})u^{\rho} = M\,, \tag{4.73}$$

where, to first order,

$$M^2 \simeq m^2 - \gamma^{\mu\nu}k_{\mu}k_{\nu} + 2k^{\rho}\Phi_{T,\rho}\,. \tag{4.74}$$

It then follows from (4.73) and (4.74) that

$$|\boldsymbol{P}| \simeq E - \frac{M^2}{2E} = E - \frac{m^2}{2E} - \frac{k^{\rho}\Phi_{T,\rho}}{E} + \frac{\gamma^{\mu\nu}k_{\mu}k_{\nu}}{2E}\,. \tag{4.75}$$

Equation (4.74) contains, as expected, the term $m^2/2$ responsible for the oscillation mechanism; the spin effects connected to Γ_{μ}; and additional gravitational contributions contained in $\Phi_{T,\rho}$, Φ_T, and in the second term on the r.h.s. of (4.67). The equation of evolution (4.67) becomes

$$i\frac{d\chi}{d\lambda} = \left(\frac{M_f^2}{2} + \Phi_G^{(f)} + \Gamma^{(f)}\right)\chi\,, \tag{4.76}$$

where the flavour mass matrix M_f is related to the vacuum mass matrix in the flavour base

$$M_f^2 = U^{\dagger} \begin{pmatrix} m_1^2 & 0 \\ 0 & m_2^2 \end{pmatrix} U\,, \tag{4.77}$$

and

$$\Phi_G^f = U^\dagger \begin{pmatrix} \langle\zeta_1|k^\rho \Phi_{G,\rho}|\zeta_1\rangle & 0 \\ 0 & \langle\zeta_2|k^\rho \Phi_{G,\rho}|\zeta_2\rangle \end{pmatrix} U\,, \tag{4.78}$$

$$\Gamma^f = U^\dagger \begin{pmatrix} \langle\zeta_1|k^\rho \Gamma_\rho|\zeta_1\rangle & 0 \\ 0 & \langle\zeta_2|k^\rho \Gamma_\rho|\zeta_2\rangle \end{pmatrix} U\,. \tag{4.79}$$

To order $O(\gamma_{\mu\nu} m_i^2/E^2)$, Eq. (4.76) becomes

$$i\frac{d\chi}{d\lambda} \simeq \left(\frac{M_f^2}{2} + k^\rho(\Phi_{G,\rho}^{(f)} + \Gamma_\rho^{(f)}) I \right) \chi\,, \tag{4.80}$$

where I is the identity matrix.

Matter effects can be easily incorporated in the diagonal element of (4.76). The effective potential V_{ν_f}, which is induced by matter and depends on neutrino flavours, is defined by [15]

$$\begin{aligned} V_{\nu_e} &= -V_{\bar{\nu}_e} = V_0(3Y_e - 1 + 4Y_{\nu_e})\,, \\ V_{\nu_{\mu,\tau}} &= -V_{\bar{\nu}_{\mu,\tau}} = V_0(Y_e - 1 + 2Y_{\nu_e})\,, \end{aligned} \tag{4.81}$$

where Y_e (Y_ν) represents the ratio between the number density of electrons (neutrinos), and

$$V_0 = \frac{G_F \rho}{\sqrt{2} m_n} = \frac{\rho}{10^{14}\mathrm{gr/cm^3}}\, 3.8\mathrm{eV}\,. \tag{4.82}$$

Here $m_n = 938\mathrm{MeV}$ is the nucleon mass, ρ is the matter density, and G_F is the Fermi coupling constant.

Transition Amplitudes

From Eqs. (4.59), (4.60) and (4.62), one obtains, to first order, the transition amplitude from a state of flavour α to one of flavour β

$$\begin{aligned} \langle\nu_\beta|\nu_\alpha(x)\rangle \simeq & \sum_{ij} U_{i\beta}^* U_{\alpha j} \langle\nu_i| e^{-i\tilde{\Phi}_T^{(j)} - ik_\mu^{(j)} x^\mu} |\nu_j\rangle + \\ & + \sum_{ij} U_{i\beta}^* U_{\alpha j} \langle\nu_i| \frac{1}{2m_j} [h_{\hat{\alpha}}^\mu \gamma^{\hat{\alpha}} (k_\mu^{(j)} + \tilde{\Phi}_{G,\,\mu}^{(j)})] e^{-ik_\beta^{(j)} x^\beta} |\nu_j\rangle\,, \end{aligned} \tag{4.83}$$

where use has been made of the relationship $\gamma^{\hat{\alpha}} k_{\hat{\alpha}}^{(j)} |\nu_j\rangle = m_j |\nu_j\rangle$. The transition probability follows from (4.83) and is defined by

$$P_{\nu_\beta \to \nu_\alpha} \equiv N |\langle\nu_\beta|\nu_\alpha(x)\rangle|^2\,, \tag{4.84}$$

where N is the normalization factor which gives the correct normalization to one of the probabilities of transition for different flavours, $\nu_\alpha \to \nu_\beta$ and $\nu_\beta \to \nu_\alpha$,

$$N \equiv \frac{1}{\langle \nu_\alpha(x)|\nu_\alpha(x)\rangle}\,, \tag{4.85}$$

and

$$|\langle \nu_\beta|\nu_\alpha(x)\rangle|^2 \simeq \sum_{ip} U^*_{i\alpha}U_{\alpha i}U_{p\beta}U^*_{p\beta}e^{-i(\tilde{\Phi}_T^{(i)}-\tilde{\Phi}_T^{(p)})-i(k^{(i)}-k^{(p)})\cdot x}+ \tag{4.86}$$

$$+\sum_{ipq} U^*_{i\beta}U_{\alpha i}U_{p\beta}U^*_{q\alpha}\langle \nu_q|\frac{1}{2m_p}[h^\mu_{\hat{\beta}}\gamma^{\hat{\beta}\,\dagger}k^{(p)}_\mu + \gamma^{\hat{\mu}\,\dagger}\tilde{\Phi}^{(p)}_{G,\mu}]|\nu_p\rangle e^{-i(k^{(i)}-k^{(p)})\cdot x}+$$

$$+\sum_{ijp} U^*_{i\beta}U_{\alpha j}U_{p\beta}U^*_{p\alpha}\langle \nu_i|\frac{1}{2m_j}[h^\mu_{\hat{\alpha}}\gamma^{\hat{\alpha}}k^{(j)}_\mu + \gamma^{\hat{\mu}}\tilde{\Phi}^{(j)}_{G,\mu}]|\nu_j\rangle e^{-i(k^{(j)}-k^{(p)})\cdot x}\,.$$

Explicit calculations are best performed by starting from (4.83) directly. With reference to the Lense–Thirring metric and (1.61) of Sect. (1.3.5), and the two flavour case ($\alpha, \beta = e, \mu$), one obtains

$$\langle \nu_e|\nu_e(x)\rangle = \sum_{i,j=1}^{2} U^*_{ie}U_{ej}O_{ij} = \cos^2\theta\, O_{11} + \sin^2\theta\, O_{22}\,, \tag{4.87}$$

where

$$O_{ij} \simeq e^{-i(k^{(i)}\cdot(x-x_0)+\Phi_G^{(i)})}\left\{1+\frac{1}{2m_i}(h^\mu_{\hat{\alpha}}k^{(i)}_\mu + \delta^\mu_{\hat{\alpha}}\Phi^{(i)}_{G,\mu})\langle \nu_i|\gamma^{\hat{\alpha}}|\nu_i\rangle - iI_{12}\right\}\delta_{ij}\,, \tag{4.88}$$

with

$$I_{1+2} = [I_1(x) + I_2(x)]\,, \tag{4.89}$$

and

$$I_1(x) \equiv \frac{1}{2}\langle \nu_i|\Phi_s|\nu_i\rangle\,,\quad I_2(x) \equiv \frac{1}{2m_i}\delta^\mu_{\hat{\alpha}}k_\mu\langle \nu_i|\gamma^{\hat{\alpha}}\Phi_s|\nu_i\rangle\,. \tag{4.90}$$

In order to separate out the real and imaginary parts of O_{ij}, it is sufficient to study $I_1 + I_2$ because the remaining terms are real. The choices $k^\mu = (E, 0, 0, k^3)$, and $k^3 \simeq E(1 - m^2/2E^2)$ are also made. Since the Lense–Thirring metric is time independent, the time-like component of the momentum E is conserved along the geodesic on which neutrinos move. Neutrino states are assumed to be eigenstates of E.

On using the relation $\nu^\dagger\gamma^{\hat{\alpha}}k_{\hat{\alpha}} = \nu^\dagger(-m+2E\gamma^{\hat{0}})$, $I_2(x)$ can be re-cast in the form

$$I_2(x) = -\frac{1}{2}\langle\nu_i|\Phi_s|\nu_i\rangle + \frac{E}{m_i}\langle\nu_i|\gamma^{\hat{0}}\Phi_s|\nu_i\rangle\,, \tag{4.91}$$

to finally obtain

$$I_1(x) + I_2(x) = \frac{E}{m_i}\langle\nu_i|\gamma^{\hat{0}}\Phi_s|\nu_i\rangle \simeq 2J_0 + O\left(\frac{m^2}{E^2}\gamma_{\mu\nu}\right)\,, \tag{4.92}$$

where

$$\begin{aligned} J_0 &= \frac{GMR^2\omega}{5}\int_{z_0}^{z}\frac{dz}{r^3}\left(1-\frac{3}{2}\frac{x^2+y^2}{r^2}\right) + \frac{GM}{4}\int_{z_0}^{z}\frac{1}{r^3}(ydx-xdy) \\ &= \frac{GMR^2\omega}{10}\frac{1}{x^2+y^2}\left[\left(\frac{z}{r}\right)^3 - \left(\frac{z_0}{r_0}\right)^3\right] + \frac{GM}{4}\frac{xy}{r}\left(\frac{1}{x^2+z^2}-\frac{1}{y^2+z^2}\right), \end{aligned} \tag{4.93}$$

and $r_0 = \sqrt{x^2+y^2+z_0^2}$. Notice that J_0 is a pure geometrical term. Equation (4.88) then reads

$$O_{ij} \simeq e^{-ik^{(i)}\cdot(x-x_0)-i\Phi_G^{(i)}}(1 + A^{(i)} + i2J_0)\delta_{ij}\,, \tag{4.94}$$

where

$$\begin{aligned} A^{(i)} \equiv A &= \frac{1}{2m_i}(h^{\mu}_{\hat{\alpha}}k^{(i)}_{\mu} + \delta^{\mu}_{\hat{\alpha}}\Phi^{(i)}_{G,\mu})\langle\nu_i|\gamma^{\hat{\alpha}}|\nu_i\rangle \\ &= \frac{1}{2}\left(h^0_{\hat{0}} + \frac{1}{E}\Phi_{G,0}\right) \\ &\simeq -\frac{1}{4}\left[\gamma_{00}(x,y,z) - \gamma_{00}(x,y,z_0)\right] - \frac{1}{4}\frac{\partial}{\partial z}\left[\int_0^x dx'\gamma_{01} + \int_0^y dy'\gamma_{02}\right] \\ &\quad + O\left(\frac{m_i^2}{E^2}\gamma_{00}\right). \end{aligned} \tag{4.95}$$

In deriving (4.95) use has been made of the relation

$$\begin{aligned} \Phi_{G,0} &\simeq \frac{E}{2}\left[\gamma_{00}(x,y,z_0) - \frac{\partial}{\partial z}\left[\int_0^x dx'\gamma_{01} + \int_0^y dy'\gamma_{02}\right] + O\left(\frac{m^2}{E^2}\gamma_{00}\right)\right] \\ &\simeq \frac{E}{2}(\gamma_{00}(z) + 4A)\,. \end{aligned} \tag{4.96}$$

Notice that to the order $\gamma_{00}m^2/E^2$, $A^{(i)} = A$ is a geometric term independent of neutrinos characteristics (mass and energy). Equation (4.87) is therefore

$$\langle\nu_e|\nu_e(x)\rangle = e^{-ik^{(1)}\cdot(x-x_0)-i\Phi_G^{(1)}}\mathcal{A}\left[\cos^2\theta + e^{-i((z-z_0)/L+\Delta\Phi_G)}\sin^2\theta\right], \tag{4.97}$$

where L and $\Delta\Phi_G$ are defined by

$$\mathcal{A} = 1 + A + i2J_0\,, \tag{4.98}$$

$$L^{-1} = \frac{\Delta m^2}{2E}\,, \quad \Delta m^2 = m_2^2 - m_1^2\,, \tag{4.99}$$

$$\Delta\Phi_G = \Phi_G^{(2)} - \Phi_G^{(1)}\,.$$

To evaluate $P_{\nu_e\to\nu_e} = N|\langle\nu_e|\nu_e(x)\rangle|^2$, one must determine the normalization factor N. By using the general expression for *spin*ors given by Eqs. (4.59) and (4.60), one finds

$$N = \frac{1}{1 + h_{\hat{0}}^0 + E^{-1}\Phi_{G,0}} \simeq 1 - (h_{\hat{0}}^0 + E^{-1}\Phi_{G,0}) \simeq 1 - 2A\,, \tag{4.100}$$

where A is given by (4.95). As it follows from (4.100), N is identical for all neutrino species, both mass eigenstates and flavour eigenstates. The flavour oscillation probability is

$$P_{\nu_e\to\nu_e} \simeq 1 - \sin^2 2\theta\, \sin^2\left(\frac{z - z_0}{2L} + \frac{\Delta\Phi_G}{2}\right). \tag{4.101}$$

In a similar way, one can find the probability for the $\nu_\mu \to \nu_e$ transition

$$P_{\nu_\mu\to\nu_e} \equiv \frac{|\langle\nu_\mu|\nu_e(x)\rangle|^2}{\langle\nu_e(x)|\nu_e(x)\rangle} \simeq \sin^2 2\theta\, \sin^2\left(\frac{z - z_0}{2L} + \frac{\Delta\Phi_G}{2}\right). \tag{4.102}$$

$\Delta\Phi_G$ must then be evaluated. From Eq. (1.71), and the fact that neutrinos propagate along null world lines so that $dz^0 \simeq dz$, one can write Φ_G as

$$\begin{aligned}\Phi_G^{(i)} &= \frac{E}{2}\left\{2\int_{z_0}^{z} dz'\gamma_{00} + G(x, y, z) + \frac{m^2}{2E^2}\left[\int_{z_0}^{z} dz'[(z - z')\gamma_{00,3} - \gamma_{00}]\right]\right\} \\ &= E\left[-(z - z_0)\gamma_{00}(z_0)\left(1 - \frac{m^2}{4E^2}\right) + \log\frac{r + z}{r_0 + z_0} + \frac{G}{2}\right],\end{aligned} \tag{4.103}$$

where

$$\begin{aligned}G(x, y, z) &= \int_0^x dx'(x - x')[\gamma_{01,1} - \gamma_{11,3}] + \int_0^y dy'(y - y')[\gamma_{02,2} - \gamma_{22,3}] \\ &= -2GMz\left[\frac{r}{y^2 + z^2} + \frac{r}{x^2 + z^2} - \left(\frac{1}{\sqrt{x^2 + z^2}} + \frac{1}{\sqrt{y^2 + z^2}}\right)\right] + \\ &+ \frac{4GMR^2\omega xy}{5}\left[-\frac{1}{y^2 + z^2}\left(\frac{1}{r} - \frac{1}{\sqrt{y^2 + z^2}}\right) + \frac{1}{x^2 + z^2}\left(\frac{1}{r} - \frac{1}{\sqrt{x^2 + z^2}}\right)\right].\end{aligned} \tag{4.104}$$

In Eq. (4.103), one should also include the quantity $EF(x, y, z)/2$, where the function F is given by

$$F = -\left(1 - \frac{m^2}{2E^2}\right)\left[\Delta x\gamma_{01} + \Delta y\gamma_{02} + 2(\Delta x\partial_1 + \Delta y\partial_2)\int dz'\gamma_{00} + \right. \tag{4.105}$$
$$\left. + \Delta z\partial_3(\int dx'\gamma_{01} + \int dy'\gamma_{02})\right]$$
$$+\Delta x\partial_1 \int dy'\gamma_{02} + \Delta y\partial_2 \int dx'\gamma_{01} + \Delta z\partial_3(\int dx'\gamma_{01} + \int dy'\gamma_{02}),$$

where $\Delta x = x - x'$, $\Delta y = y - y'$, and $\Delta z = z - z'$. F can be made to vanish by choosing $x' = x$, $y' = y$, and $z' = z$. In order to compare (4.103) with the results obtained with different approaches, it is convenient to consider the *radial motion*, for which one can take, without loss of generality, $x \sim y \sim 0$, i.e. $G = 0$. Equation (4.103) then becomes

$$\Phi_G^{(i)} \simeq \frac{E}{2}\left[2\int_{z_0}^{z} dz'\gamma_{00} + \frac{m^2}{2E^2}\left(\int_{z_0}^{z} dz'(z - z')\gamma_{00,3} - \int_{z_0}^{z} dz'\gamma_{00}\right)\right], \tag{4.106}$$

from which one derives

$$\Delta\Phi_G = -\frac{\Delta m^2}{4E}[(z - z_0)\gamma_{00}(z_0)]. \tag{4.107}$$

The phase in (4.101) becomes

$$\Omega \equiv \frac{z - z_0}{2L} + \frac{\Delta\Phi_G}{2} \sim \frac{z - z_0}{2L}, \tag{4.108}$$

where the constant factor $1 - \gamma_{00}(z_0)/2 \simeq (g_{00}(z_0))^{-1/2}$ has been absorbed in a redefinition of the constant $E \to E/\sqrt{g_{00}(z_0)}$. The coordinate difference $z - z_0$ is not the physical distance, the latter being defined as

$$l = \int_{z_0}^{z} \sqrt{-g_{33}}\, dz' \sim \int_{z_0}^{z} (1 - \gamma_{00}/2)dz'.$$

Notice also that E is the neutrino energy measured by an inertial observer at rest at infinity. Besides, the phase Ω can be rewritten in terms of physical quantities by introducing $E_l = E/\sqrt{g_{00}(z)}$, which is the energy measured by a locally inertial observer momentarily at rest in the gravitational field. Since $z - z_0 = \int_{l_0}^{l} dl/\sqrt{-g_{33}}$ it follows that, in WFA,

$$\Omega = \frac{\Delta m^2}{4E}(z - z_0) \simeq \frac{\Delta m^2}{4}\int_{l_0}^{l}\frac{dl}{E_l}, \tag{4.109}$$

which reflects the fact that the space–time curvature affects the oscillation probability through the gravitational redshift of the local energy E_l and the proper distance dl. These results are consistent with those of Refs. [16,17].

Quantum Mechanical Phase: Independence of γ_{0i}

As it follows from (4.107), $\Delta\Phi_G$ does not depend on γ_{0i}, which only occurs in G. A similar result follows from the definition of momentum by means of Eq. (4.118). In fact, choosing $k^\mu = (E, 0, 0, k^3)$ one gets

$$P_3 = \frac{\partial \upsilon}{\partial x^3} = -k_3 + \frac{E}{2}(\gamma_{33} - \gamma_{00}) . \tag{4.110}$$

Since for the Lense–Thirring metric $\gamma_{00} = \gamma_{33}$, the gravitational contribution vanishes, and one has $P_3 = -k_3$. The phase υ reads

$$\upsilon = -E\left(1 - \frac{m^2}{2E^2}\right)(z - z_0) + \frac{E}{2}\int_{z_0}^{z} dz'(\gamma_{33} - \gamma_{00}) , \tag{4.111}$$

so that $\Delta\upsilon$, which is the quantity entering into the quantum mechanical phase of neutrinos, is given by

$$\Delta\upsilon = \frac{\Delta m^2}{2E}(z - z_0) , \tag{4.112}$$

i.e. $\Delta\upsilon = \Omega$ (see Eq. (4.108)). The fact that γ_{0i} does not appear in the expression for Ω or $\Delta\upsilon$ means that these quantities depend quadratically (and *not* linearly) on the angular momentum of the neutrinos. This result is expected because, as noted by Wudka [17], the quantum mechanical phase is a scalar whereas the angular momentum is a pseudovector.

Estimate of the Coefficients C

Consider again the expression for the matrix C defined in (4.69). After some algebras, the element of the matrices C_{ij} can be rewritten as

$$C_{ij}^{(i)} \sim -i\frac{E}{4}\gamma_{00,3}e^{-i(k^{(i)}\cdot(x-x_0)-\Phi_G^{(i)})}\delta_{ij} . \tag{4.113}$$

The expression for O_{ij} derived in (4.94) can be used to re-cast (4.113) into the form

$$C_{ij}^{(i)} = C_{ij} \sim -i\frac{E}{4}\frac{\gamma_{00,3}}{1 + A + i2J_0}O_{ij} = \bar{C}O_{ij} , \tag{4.114}$$

where

$$\bar{C} \equiv -i\frac{E}{4}\frac{\gamma_{00,3}}{1 + A + i2J_0} \sim , -i\frac{E}{4}\gamma_{00,3} \tag{4.115}$$

and A and J_0 are defined in Eqs. (4.95) and (4.93), respectively. Notice that to order $O(m^2/E^2)$, $\bar{C}$ does not depend on the mass eigenstate index (i). Equation (4.68) therefore becomes

$$i\,\frac{dO_{ij}}{d\lambda} \backsimeq A^{(i)}\,O_{ij} + \sum_k d^{(i)}_{ik}\,O_{kj} + \bar{C}O_{ij}\,. \tag{4.116}$$

The matrix C in Eq. (4.69) then is

$$C = U^\dagger \begin{pmatrix} \bar{C}O_{11} & 0 \\ 0 & \bar{C}O_{22} \end{pmatrix} U\phi = \bar{C}\chi\,. \tag{4.117}$$

The dependence on $\bar{C}$ in the equation of evolution (4.70) can be removed by the transformation $\chi \to e^{-i\int_{\lambda_0}^{\lambda}\bar{C}d\lambda}\chi$ and then by a choice of normalization.

4.5 Neutrino Lensing

Once $\Psi_0(x)$ is chosen to be a plane-wave solution of the flat space–time Dirac equation, the phase of a mass eigenstate neutrino is given by

$$\upsilon(x) = -k_\alpha x^\alpha - \tilde{\Phi}_G\,. \tag{4.118}$$

Neglecting background matter effects, the geometrical optics of the neutrino in the presence of gravitational fields can be determined from

$$\begin{aligned} P_\mu &= \frac{\partial \upsilon}{\partial x^\mu} = -k_\mu - \Phi_{G,\mu}\,, \\ &= -k_\mu - \frac{1}{2}\,\gamma_{\alpha\mu}(x)k^\alpha + \frac{1}{2}\int_P^x dz^\lambda(\gamma_{\mu\lambda,\beta}(z) - \gamma_{\beta\lambda,\mu}(z))k^\beta\,. \end{aligned} \tag{4.119}$$

The space components

$$\frac{\partial \upsilon}{\partial x^i} = -k_i - \frac{1}{2}\,\gamma_{\alpha i}k^\alpha + \frac{1}{2}\int_P^x dz^\lambda(\gamma_{i\lambda,\beta} - \gamma_{\beta\lambda,i})k^\beta \tag{4.120}$$

give the direction of propagation of the neutrinos in a gravitational background, while the index of refraction is

$$n = \left(\frac{\partial\phi}{\partial x^0}\right)^{-1}\sqrt{\left(\frac{\partial\phi}{\partial x^1}\right)^2 + \left(\frac{\partial\phi}{\partial x^2}\right)^2 + \left(\frac{\partial\phi}{\partial x^3}\right)^2}\,. \tag{4.121}$$

Equations (4.121) and (4.120) are now applied to the Lense–Thirring metric (1.59). The contributions to $\partial \upsilon/\partial x^i$ can be promptly derived and are

$$\frac{\partial \phi}{\partial x^i} = a_i - \frac{1}{2}\gamma_{0i}k^0 + \frac{1}{2}\int_P^x dz^0(\gamma_{i0,j}k^j - \gamma_{j0,i}k^j) - \frac{1}{2}\int_P^x dz^j \gamma_{0j,i}k^0 , \tag{4.122}$$

where

$$a_i = -k_i - \frac{1}{2}\gamma_{ij}k^j + \frac{1}{2}\int_P^x dz^j(\gamma_{ij,\beta}k^\beta - \gamma_{lj,i}k^l) - \frac{1}{2}\int_P^x dz^0 \gamma_{00,i}k^0 . \tag{4.123}$$

z-**axis propagation**. For simplicity, the contributions of order m^2 are neglected. Then $z^0 \cong z$ and the integrals that contain γ_{0l} in (4.122) vanish, while for $z \gg R$, in the neighbourhood of the source ($x = R\cos\alpha$, $y = R\sin\alpha$) and one finds ($k^3 \equiv k$)

$$\frac{1}{2}\gamma_{01}k^0 \simeq \frac{1}{2}\gamma_{01}k^3 \simeq k\,\frac{2GM\omega}{5}\left(\frac{R}{z}\right)^3 \sin\alpha\,, \tag{4.124}$$

$$\frac{1}{2}\gamma_{02}k^0 \simeq \frac{1}{2}\gamma_{02}k^3 \simeq -k\,\frac{2GM\omega}{5}\left(\frac{R}{z}\right)^3 \cos\alpha\,.$$

Finally, the quantities a_i must be evaluated from (4.122). For large values of z, one finds

$$\begin{aligned}
a_1 &= -k\left(\int_{-z_0}^{z_0} dz^0\phi_{,1} + \int_{-z}^{z} dz^3\phi_{,1}\right) \simeq -k\,\frac{2GM}{R}\cos\alpha\left(1+\frac{z}{r}\right),\\
a_2 &= -k\left(\int_{-z_0}^{z_0} dz^0\phi_{,2} + \int_{-z}^{z} dz^3\phi_{,2}\right) \simeq -k\,\frac{2GM}{R}\sin\alpha\left(1+\frac{z}{r}\right),\\
a_3 &= -k(1+\phi),
\end{aligned} \tag{4.125}$$

where z^0 is the time at which the neutrino is at z. Though the path followed by the neutrinos is not in general plane, its deflection from the original non-perturbed direction can be calculated from the expression

$$\tan\varphi = \frac{1}{k^3}\sqrt{\left(\frac{\partial \upsilon}{\partial x^1}\right)^2 + \left(\frac{\partial \upsilon}{\partial x^2}\right)^2} = \frac{4GM}{R}\,, \tag{4.126}$$

which coincides with the result of General Relativity. The result also agrees with the results of neutrino optics where Gaussian wave packets have been used.

4.6 Spin-Gravity Coupling Of Neutrinos with Primordial Gravitational Waves

The approach discussed in this section is based on quantum field theory in curved space–time and has been proposed in [18].

It regards the spin-gravity coupling of neutrinos with primordial gravitational waves. The latter are generated during the initial inflationary phase of the Universe.

Using the properties of the Dirac matrices, the Lagrangian density can be rewritten in the form

$$\mathcal{L} = \det(e)\, \bar{\psi} \left(i\gamma^a \partial_a \;-\; m \;-\; \gamma_5 \gamma_d B^d \right) \psi,$$

where

$$B^d = \epsilon^{abcd} e_{b\lambda} (\partial_a e_c^\lambda + \Gamma^\lambda_{\alpha\mu} e_c^\alpha e_a^\mu).$$

In a local inertial frame of the fermion, the effect of a gravitational field appears as the *axial-vector interaction* term shown in $\mathcal{L}$.

The quantity B^d is calculated using the perturbed Friedmann–Robertson–Walker Universe, where perturbations are generated by quantum fluctuations of inflatons. Because for a Universe of this type the B^a-term vanishes due to the symmetry of the metric, the general form of perturbations on a flat Robertson–Walker expanding universe can be written as

$$ds^2 = a(\tau)^2 [(1 + 2\Phi) d\tau^2 - \omega_i dx^i d\tau - ((1 + 2\Psi)\delta_{ij} + h_{ij}) dx^i dx^j],$$

where Φ, Ψ are scalar fluctuations, ω_i is the vector fluctuations, and h_{ij} is the tensor fluctuations of the metric. Of the ten degrees of freedom in the metric perturbations only six are independent and the remaining four can be set to zero by suitable gauge choices. In the transverse-traceless gauge, $h^i_i = 0$, $\partial^i h_{ij} = 0$, and the perturbed metric can be expressed as

$$\begin{aligned} ds^2 &= a(\tau)^2 [(1 + 2\Phi) d\tau^2 - \omega_i dx^i d\tau - (1 + 2\Psi - h_+) dx_1^2 \\ &\quad -(1 + 2\Psi + h_+) dx_2^2 - 2h_\times dx_1 dx_2 - (1 + 2\Psi) dx_3^2]. \end{aligned} \tag{4.127}$$

An orthogonal set of vierbiens e^a_μ for this metric is given by

$$e^a_\mu = a(\tau) \begin{pmatrix} 1+\phi & -\omega_1 & -\omega_2 & -\omega_3 \\ 0 & -(1+\psi) + h_+/2 & h_\times & 0 \\ 0 & 0 & -(1+\psi) - h_+/2 & 0 \\ 0 & 0 & 0 & -(1+\psi)\,. \end{pmatrix}$$

Only the tensor perturbations are needed, therefore $B^a = (\partial_\tau h_\times, 0, 0, \partial_\tau h_\times)$. This affects the dispersion relations of neutrinos. Consequences of these results in the framework of leptogenesis have been studied in [18,19].

4.7 Pulsar Kick

In this section, the role of spin-gravity coupling is discussed in an astrophysical context.

Consider the spin-flip conversion of neutrinos propagating in the gravitational field of a rotating source. The geometry of a rotating mass is described by the Lense–Thirring line element.

The covariant Dirac equation in curved space–time is rewritten in an appropriate form. Using the equation $[i\gamma^\mu(x)D_\mu - m]\psi = 0$, and the relation

$$\gamma^{\hat{a}}[\gamma^{\hat{b}}, \gamma^{\hat{c}}] = 2\eta^{\hat{a}\hat{b}}\gamma^{\hat{c}} - 2\eta^{\hat{a}\hat{c}}\gamma^{\hat{b}} - 2i\epsilon^{\hat{d}\hat{a}\hat{b}\hat{c}}\gamma_{\hat{d}}\gamma^5\,, \tag{4.128}$$

where $\epsilon^{\hat{d}^a\hat{b}^c}$ is the totally antisymmetric Ricci tensor ($\epsilon^{0123} = +1$), the spin connection can be cast into the convenient form [16]

$$\gamma^\mu\Gamma_\mu = \gamma^{\hat{a}}e^\mu_{\hat{a}}\left\{iA_{G\mu}\left[-(-g)^{-1/2}\gamma^5\right]\right\} \tag{4.129}$$

where

$$A^\mu_G = \frac{1}{4}\sqrt{-g}e^\mu_{\hat{a}}\varepsilon^{\hat{d}\hat{a}\hat{b}\hat{c}}(e_{\hat{b}\nu;\sigma} - e_{\hat{b}\sigma;\nu})e^\nu_{\hat{c}}e^\sigma_{\hat{d}}\,, \tag{4.130}$$

and $g = det(g_{\mu\nu})$. The corresponding non-vanishing components of the vierbein are $e_{0\hat{0}} = (1+\phi)$, $e_{0\hat{i}} = -h^i$, $e_{j\hat{i}} = -(1-\phi)\delta_{ji}$, so that A^μ_G becomes

$$A^\mu_G(x) = \left(0, -\frac{4}{5}\frac{GMR_s^2}{r^3}\boldsymbol{\omega}'\right)\,, \quad \boldsymbol{\omega}' = \boldsymbol{\omega} - \frac{3(\boldsymbol{\omega}\cdot\boldsymbol{x})\boldsymbol{x}}{r^2}\,.$$

For a Schwarzschild space–time, characterized by spherical symmetry, A^μ_G vanishes. The term $\gamma^\mu\Gamma_\mu \sim \gamma^\mu A_{G\,\mu}\gamma^5$ implies modified dispersion relations for left- and right-handed neutrino states. In fact, writing $\gamma^5 = \mathcal{P}_R - \mathcal{P}_L$ ($\mathcal{P}_{L,R} = (1 \mp \gamma^5)/2$), it follows that left- and right-handed neutrinos acquire gravitational potentials of opposite signs. The dispersion relations of ν_L's (an arbitrary right-handed (SU(2)-singlet)) and ν_R's, $\eta^{\hat{a}\hat{b}}(p_{\hat{a}} + \xi A_{G\,\hat{a}})(p_{\hat{b}} + \xi A_{G\,\hat{b}}) = m^2$ ($\xi = -1$ for ν_L and $\xi = 1$ for ν_R), are therefore shifted by terms $\sim +\boldsymbol{p}\cdot\boldsymbol{A}_G$ and $\sim -\boldsymbol{p}\cdot\boldsymbol{A}_G$, respectively. Therefore, one gets $E_{L,R}(p) = p + \frac{m^2}{2p} \pm \frac{\boldsymbol{p}\cdot\boldsymbol{A}_G}{p}$.

These results can have interesting consequences for the pulsar "kick". Consider electron neutrinos emitted by a nascent protostar. The ν_e's inside their neutrinospheres are trapped owing to weak interactions with the matter of the background, which leads to the potential energy

$$V_{\nu_e}(r) \simeq 3.8\,\frac{\rho}{10^{14}\text{gr cm}^{-3}}\,y_e(r)\,\text{eV}\,, \quad y_e = Y_e - \frac{1}{3}\,, \tag{4.131}$$

where ρ is the matter density and Y_e is the electron fraction. Moreover, electron neutrinos interact with the (uniform) magnetic fields of the protostars via

$\mathcal{L}_{int} = \bar{\nu}\hat{\mu}\sigma^{\hat{a}\hat{b}}F_{\hat{a}\hat{b}}\nu$, where $\hat{\mu}$ is the magnetic momentum of the neutrino, $F_{\hat{a}\hat{b}}$ is the electromagnetic field tensor, and $\sigma^{\hat{a}\hat{b}} = \frac{i}{2}[\gamma^{\hat{a}}, \gamma^{\hat{b}}]$. The equation of evolution which describes the spin-flip conversion then is [20]

$$i\frac{d}{dr}\begin{pmatrix}\nu_{eL}\\ \nu_{eR}\end{pmatrix} = \mathcal{H}\begin{pmatrix}\nu_{eL}\\ \nu_{eR}\end{pmatrix}, \tag{4.132}$$

where the matrix $\mathcal{H}$ is the effective Hamiltonian

$$\mathcal{H} = \begin{bmatrix} V_{\nu_e} + \mathbf{\Omega}_G \cdot \hat{p} & \mu B_\perp^* \\ \mu B_\perp & -\mathbf{\Omega}_G \cdot \hat{p} \end{bmatrix}, \tag{4.133}$$

$$\mathbf{\Omega}_G(r) = \frac{4GMR_s^2}{5r^3}\omega' \simeq 8\ 10^{-13}\frac{M}{M_\odot}\left(\frac{R_s}{10\text{km}}\right)^2\left(\frac{10\text{km}}{r}\right)^3\frac{\omega'\cos\phi}{10^4\text{Hz}}\text{eV}. \tag{4.134}$$

$B_\perp = B\sin\alpha$ is the component of the magnetic field orthogonal to the neutrino momentum, and ϕ is the angle between the neutrino momentum and the angular velocity. The quantity $\mathbf{\Omega}_G$ is diagonal in spin space and cannot, therefore, induce spin-flips. Its relevance comes from the fact that it modifies the resonance conditions of the transition $\nu_{eL} \to \nu_{eR}$

$$V_{\nu_e}(\bar{r}) + 2\mathbf{\Omega}_G(\bar{r}) \cdot \hat{p} = 0\,, \tag{4.135}$$

where $\bar{r}$ is the resonance point. $\mathbf{\Omega}_G$ in (4.135) distorts the surface of resonance due to the relative orientation of the neutrino momentum with respect to the angular velocity. As a consequence, the outgoing energy flux is modified and generates the pulsar kick. Such a result is similar to that proposed in [21] where the coupling of neutrinos with magnetic fields was considered. More precisely, the mechanism relies on the fact that a resonant oscillation $\nu_e \to \nu_{\mu,\tau}$ may occur between the ν_e and $\nu_{\mu,\tau}$ neutrinospheres. The neutrinos ν_e are trapped by the medium, but neutrinos $\nu_{\mu,\tau}$ generated by oscillations may escape from the protostar because are outside their neutrinosphere: the "surface of resonance" acts as an "effective muon/tau neutrinosphere". Moreover, the magnetic field of the protostar deforms the surface of resonance, i.e. the usual MSW resonance conditions turn out to be modified by the term [22] $\mathbf{\Omega}_B \cdot \hat{p}$, where $\mathbf{\Omega}_B = \frac{eG_F}{\sqrt{2}}\sqrt[3]{\frac{3n_e}{\pi^4}}\,\mathbf{B}$, $\hat{p} = \mathbf{p}/p$, $\mathbf{p}$ is the neutrino momentum, G_F is the Fermi constant, and n_e is the electron density. The discussion of sterile neutrinos is given in [23,24].

Problems

4.1 Calculate the change in phase produced by Earth's rotation in a neutrino beam produced at CERN and detected at the Gran Sasso Laboratory (distance ℓ). Complete the space–time path of this interferometer by assuming that the neutrinos at the source can be transported coherently to the Gran Sasso.

4.2 Calculate the effect of the gravitational field of our Sun on a photon interferometer. Treat the Sun as a solid sphere.

4.3 Repeat the same calculation of Problem 4.2 for a superconducting interferometer.

4.4 Problem Heading

4.1 Consider relativistic neutrinos of energy $k^0 \simeq k$. Neglect spin at first. Then using Expression (1.38) of Chap. 1 for Φ_G leads to the result $\Delta\chi \sim k\omega\ell^2$. This result can be re-obtained by observing that at any instant in time the neutrinos of energy k of a length dr of beam are subject to the flux of a gravitational field $d\Phi = k\omega r dr$. The spin contribution can be calculated by substituting the results of Sect. 1.3.8 in $\phi_s = \int^{\ell} dz^{\lambda}\Gamma_{\lambda}$.
4.2 The gravitational field of a rotating solid sphere is given in Sect. 3.2.2 where M and R now refer to our Sun and the system of coordinates r, θ, φ is centred at the Sun. The part of χ that does not contain spin is similar to that derived in Problem 4.1, with the appropriate replacement for the energy of the neutrino. The spin part of the calculation is summarized in Sect. 1.3.6 and is $\xi = \int dt \mathbf{\Omega} \cdot dS$.
4.3 Assume that the superconductor is relativistic. One can apply in this case the solution of the Klein–Gordon equation and calculate Φ_G as in Sect. 3.2.1.

References

1. Wolfenstein, L.: Phys. Rev. D **17**, 2369 (1978)
2. S.P. Mikheyev and A. Yu. Smirnov, Sov. J. Nucl. Phys. **42**, 913 (1985); Nuovo Cimento C **9**, 17 (1986)
3. Y.Q. Cai, G. Papini, Phys. Rev. Lett. **66**, 1259 (1991); G. Papini, *Proc of the* 5th *Canadian Conference on General Relativity and Relativistic Astrophysics*, University of Waterloo 13-15 May 1993, edited by R.B. Mann and R.G. McLenaghan, World Scientific, Singapore 1994, p. 10
4. V. de Sabbata and C. Sivaram, Nuovo Cimento A **104**, 1577 (1991). These author consider the possibility that large magnetic fields exist in the innermost region of the Sun's core
5. V. de Sabbata and C. Sivaram, Nuovo Cimento B **105**, 1181 (1990). The authors show that higher magnetic moments may occur as a consequence of the Kaluza-Klein theory
6. Dvornikov, M.: arXiv: hep-ph/0601095 v2 (2006); Phys. J. C **80**, 474 (2020)
7. Ahluwalia, D.V., Burgard, C.: Phys. Rev. D **57**, 4274 (1998)
8. Bhattacharya, T., Habib, S., Mottola, E.: Phys. Rev. D **59**, 067301 (1999)
9. de Sabbata, V., Gasperini, M.: Nuovo Cimento A **65**, 479 (1981)
10. de Sabbata, V., Sivaram, C.: Annalen der Physik Leipzig **46**, 8 (1989)
11. S. Chakraborty, JCAP **10**, 019 (2015). S. Capozziello, G. Lambiase, Mod. Phys. Lett. A **14**, 2193 (1999). L. Buoninfante, G.G. Luciano, L. Petruzziello, L. Smaldone, Phys. Rev. D **101**, 024016 (2020)
12. Weinberg, S.W.: Gravitation and Cosmology. Wiley, New York (1972)
13. J. Lense and H. Thirring, Z. Phys. **19**, 156 (1918); (English translation: B. Mashhoon, F.W. Hehl and D.S. Theiss, Gen. Rel. Grav. **16**, 711(1984))

14. Lambiase, G., Papini, G., Punzi, R., Scarpetta, G.: Phys. Rev. D **71**, 073011 (2005)
15. Bilenky, S.M., Giunti, C., Grimus, W.: Prog. Part. Nucl. Phys. **43**, 1 (1999)
16. Cardall, C.Y., Fuller, G.M.: Phys. Rev. D **55**, 7960 (1997)
17. Wudka, J.: Phys. Rev. D **64**, 065009 (2001)
18. Lambiase, G., Mohanty, S., Prasanna, A.R.: Phys. Rev. Lett. **96**, 071302 (2006)
19. S. Mohanty, B. Mukhopadhyay, A.R. Prasanna, hep-ph/0204257 [hep-ph]. M. Sinha and B. Mukhopadhya, Phys. Rev. D **77**, 025003 (2008). B. Mukhopadhya, Class. Quant. Grav **25**, 065006 (2008). U. Debnath, B. Mukhopadhya, and N. Dadhich, Mod. Phys. Lett. A **21**, 399 (2006)
20. M.B. Voloshin, Phys. Lett. B **209**, 360 (1988); JEPT Lett. **47**, 501 (1988)
21. Kusenko, A., Segré, G.: Phys. Rev. Lett. **77**, 4872 (1996)
22. J.F. Nieves, P.B. Pal, Phys. Rev. D **40**, 1693 (1989). S. Esposito, S. Capone, Z. Phys. C **70**, 55 (1996). J. C. D'Olivo, J.F. Nieves, P.B. Pal, Phys. Rev. D **40**, 3679 (1989); Phys. Rev. Lett. **64**, 1088 (1990). J.C. D'Olivo, J.F. Nieves, Phys. Lett. B **683**, 87 (1996). P. Elmfors, D. Grasso, G.G. Raffelt, Nucl. Phys. B **479**, 3 (1996)
23. Kusenko, A., Segré, G.: Phys. Lett. B **396**, 197 (1997)
24. Fuller, G.M., Kusenko, A., Mocioiu, I., Pascoli, S.: Phys. Rev. D **68**, 103002 (2004)

Neutrinos Physics: Further Topics

5

Abstract

This chapter deals with general relativistic effects on spin-flavour oscillations above the core of type II supernovae. The evolution equation is derived and the relative magnitudes of the terms in the Hamiltonian are discussed. These terms arise from weak, electromagnetic, and gravitational interactions. The effects on resonance position and adiabaticity are studied. A semiclassical treatment of spin-flip is presented.

5.1 Introduction—The Standard Model

The Standard Model of particle physics is the theory that unifies the electromagnetic weak and strong interactions (see Fig. 5.1).

The electromagnetic interactions describe the interactions of charged particles. The relativistic quantum theory that describes this interaction is Quantum Electrodynamics, which is a $U_Q(1)$ gauge theory (Q corresponds to the electric charge). The mediator of the electromagnetic interaction is the photon that represents the gauge boson corresponding to the generator of the $U_Q(1)$ group. The weak interaction describes β-decays. At low energies the Lagrangian density describing this interaction is given by an effective four fermion interaction represented by an operator of dimension 6. The coupling constant has therefore dimensions of an inverse mass-squared. At high energies, this effective theory fails to correctly describe the weak interactions, and one must consider the electro-weak interaction, which is a renormalizable theory that unifies both weak and electromagnetic interactions. The electro-weak theory is described by the $SU_L(2) \times U_Y(1)$ group, which is spontaneously broken down to $U_Q(1)$ at energies ~ 100 GeV. The spontaneous symmetry breaking enables the generation of the masses of three gauge bosons, $W^{\pm}$, Z, leaving massless only the photon. Under the $SU_L(2)$ group, the left-handed leptons l_L and quarks q_L can have three colours and six flavours $i = u, d, c, s, t, b$, while the right-handed fields l_R, q_R are singlets. This implies that the mass terms for the fermions

G. Lambiase and G. Papini, *The Interaction of Spin with Gravity in Particle Physics*, Lecture Notes in Physics 993, https://doi.org/10.1007/978-3-030-84771-5_5

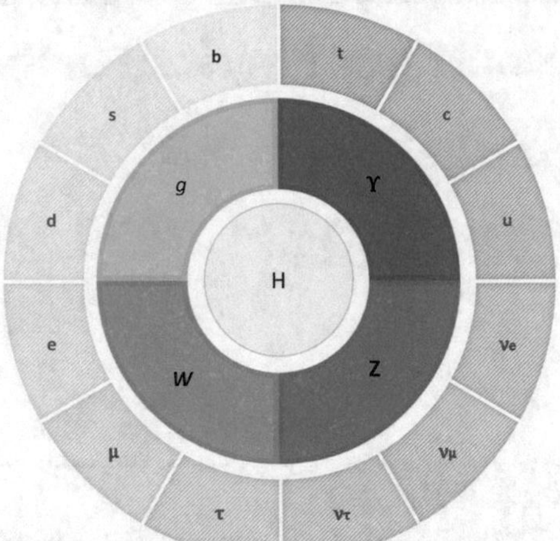

Fig. 5.1 The Standard Model particles in a nutshell: quarks, leptons, gauge bosons, and the Higgs boson

are forbidden before the electro-weak symmetry breaking. After the electro-weak phase transition, the Higgs scalar acquires a non-vanishing expectation value (vev) that breaks the electro-weak symmetry, so that the quarks and leptons can acquire a mass. The problem of neutrino mass is somewhat different and will be briefly discussed later. The strong interaction, finally, is the interaction among the quarks of different colours. The interaction is mediated by eight gluons. The underlying group is $SU(3)$, and the corresponding gauge theory describing quarks and their interaction is called Quantum Chromodynamics (QCD). The three coloured states of every flavour belong to the triplet representation of the $SU_c(3)$, $\Psi^i_\alpha = \begin{pmatrix} q^i_1 \\ q^i_2 \\ q^i_k \end{pmatrix}$, where $\alpha = r, g, b$ is the colour index, and i is the flavour index, while the gluons are the eight gauge bosons, G^a_μ, $a = 1, ..., 8$, and correspond to the generators of the group $SU_c(3)$. Coloured states are confined (confinement of quarks). Only colour singlet states can exist in nature as free particles, which are baryons (made of three quarks), i.e. $\varepsilon^{\alpha\beta\rho} q^i_\alpha q^j_\beta q^k_\rho$, or mesons (made of quark–antiquark pairs), i.e. $\delta^\alpha_\beta \bar{q}^{\beta i} q^j_\alpha$. The strong nuclear force describes the interaction between the protons and neutrons. It is a manifestation of the underlying $SU_c(3)$ interactions among the quarks. In the SM, hence, the symmetry group is given by $SU_c(3) \times SU_L(2) \times U_Y(1)$.

The quarks (and leptons) are assigned to be left-handed doublets and right-handed singlets. However, the quark mass eigenstates (d, s, b) are not the same as the weak eigenstates (d', s', b'). The matrix relating these bases is the 3×3 Cabibbo, Kobayashi, Maskawa (CKM) matrix (V_{CKM})

$$\begin{pmatrix} d' \\ s' \\ b' \end{pmatrix} = \begin{pmatrix} V_{ud} & V_{us} & V_{ut} \\ V_{cd} & V_{cs} & V_{cb} \\ V_{td} & V_{ts} & V_{tb} \end{pmatrix} \begin{pmatrix} d \\ s \\ b \end{pmatrix} . \tag{5.1}$$

The V_{CKM} matrix can be parameterized in terms of three mixing angles and the CP-violating phase, namely,

$$V_{CKM} = \begin{pmatrix} c_{12}c_{13} & s_{12}c_{13} & s_{13}e^{i\delta} \\ -s_{12}c_{23} - c_{12}s_{23}s_{13}e^{i\delta} & c_{12}c_{23} - s_{12}s_{23}s_{13}e^{i\delta} & s_{23}c_{13} \\ s_{12}s_{23} - c_{12}c_{23}s_{13}e^{i\delta} & -c_{12}s_{23} - s_{12}c_{23}s_{13}e^{i\delta} & c_{23}c_{13} \end{pmatrix}, \tag{5.2}$$

with $s_{ij} = \sin\theta_{ij}$, $c_{ij} = \cos\theta_{ij}$, and

$$s_{12} = \frac{|V_{us}|}{\sqrt{|V_{ud}|^2 + |V_{us}|^2}}, \quad s_{23} = \left|\frac{V_{cb}}{V_{us}}\right|, \quad s_{13}e^{i\delta} = V_{ub}^*.$$

As mentioned, the weak interactions are mediated in the SM by the exchange of the gauge bosons $W^{\pm}$, Z. In the case of $W^{\pm}$, the flavours of the quarks interacting with the gauge boson can change, namely $W^{\pm}$ couples to quark pairs (u, d), (c, s), (t, b), and to leptons (ν_e, e^-), (ν_μ, μ^-), (ν_τ, τ^-). It follows that particles containing strange quarks s, such as $K^{\pm}$, K^0, Λ, etc., that cannot decay, via strong interactions, into hadrons that do not contain a strange quark (the flavour must be conserved), can however decay via the weak interactions. This is possible because $W^{\pm}$ not only couples to u, d quarks, but also (with a weak coupling g_W) to u, s quarks, i.e. usW^+, with a vertex $g_W \sin\theta_C$, while the vertex udW is associated to the coupling $g_W \cos\theta_C$. Here $\theta_C \approx 0.22$ is the Cabibbo angle. This coupling allows, for example, the decay $\Lambda \to p + e^- + \bar{\nu}_e$, which takes place since an s-quark converts into a u-quark and emits a gauge boson W^-, which then decays into an electron e^- and anti-neutrino $\bar{\nu}_e$ (see Fig. 5.2). Similarly, the c-quark couples to the s-quark with coupling $g_W \cos\theta_C$, as well as to the d-quark with coupling $-g_W \sin\theta_C$. In the case of two flavours, the mixing field (5.1) is

$$\begin{pmatrix} u \\ c \end{pmatrix} = \begin{pmatrix} \cos\theta_C & \sin\theta_C \\ -\sin\theta_C & \cos\theta_C \end{pmatrix} \begin{pmatrix} u \\ c \end{pmatrix}. \tag{5.3}$$

In the SM, neutrinos are massless particles. An extension of the SM includes the introduction of massive neutrinos, $L_{\nu-mass} = m_i \nu_{iL} \nu_{iL}$, where ν_{iL} are the physical

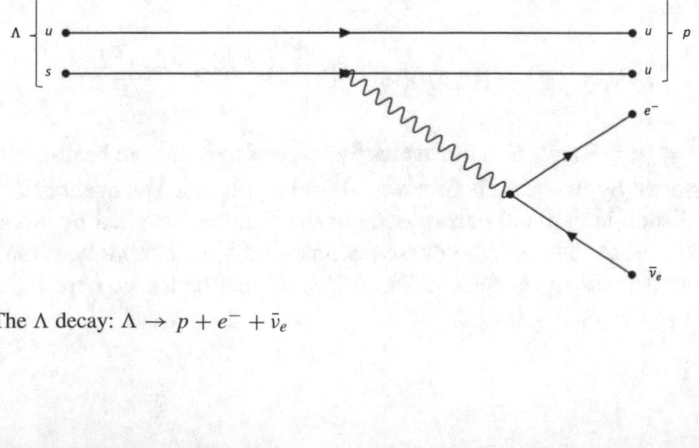

Fig. 5.2 The Λ decay: $\Lambda \to p + e^- + \bar{\nu}_e$

states with (Majorana) mass m_i. As for the quarks, the physical states are not in general the weak eigenstates that appear in the charged current and neutral current interactions of the neutrinos. The weak eigenstates $\nu_{\alpha L}$ are related to the physical or mass eigenstates ν_{iL} by

$$\nu_{\alpha L} = \sum_{i=1}^{3} U_{\alpha i} \nu_{iL} , \quad \alpha = e, \mu, \tau , \tag{5.4}$$

where $U_{\alpha i}$ is the Pontecorvo–Maki–Nakagawa–Sakata (PMNS) mixing matrix. The mass matrix in the flavour basis (the basis in which the charged lepton mass matrix is diagonal) can be written as $M^{(\nu)}_{\alpha\beta} = U_{\alpha i} M^{diag}_{ij} U^T_{\beta j}$, where M^{diag}_{ij} is the diagonal matrix mass $M^{diag}_{ij} = m_i \delta_{ij}$. The mixing matrix $U_{\alpha i}$ is parameterized in terms of three mixing angles,

$$U = \begin{pmatrix} c_1 c_3 & -s_1 c_3 & -s_3 \\ s_1 c_2 - c_1 c_2 s_3 & c_1 c_2 + s_1 s_2 s_3 & -c_3 s_1 \\ s_1 s_2 + c_1 c_2 s_3 & c_1 s_2 - s_2 c_2 s_3 & c_2 c_3 \end{pmatrix} , \tag{5.5}$$

with $s_i = \sin\theta_i$ and $c_i = \cos\theta_i$. In the presence of CP violation, a complex phase appears in the matrix U.

Consider now the neutrino oscillations for two neutrino flavours, with weak eigenstates $|\nu_e\rangle$ and $|\nu_\mu\rangle$. The mass matrix in this basis is not diagonal. After the diagonalization of the mass matrix, the two physical eigenstates are given by $|\nu_1\rangle$ and $|\nu_2\rangle$, with masses m_1 and m_2. The two weak (flavour) eigenstates can now be written in terms of the mass eigenstates as

$$\begin{pmatrix} |\nu_e\rangle \\ |\nu_\mu\rangle \end{pmatrix} = \begin{pmatrix} \cos\theta & \sin\theta \\ -\sin\theta & \cos\theta \end{pmatrix} \begin{pmatrix} |\nu_1\rangle \\ |\nu_2\rangle \end{pmatrix} . \tag{5.6}$$

The physical states evolve with time as $|\nu_{1,2}(t)\rangle = e^{-iE_{1,2}t}\, |\nu_{1,2}\rangle(t = 0)$. If initially only an electron neutrino $|\nu_e\rangle$ appears in the beam, after a time t both $|\nu_e\rangle$ and $|\nu_\mu\rangle$ will be present in the beam. For the ultra-relativistic neutrinos with momentum $p \gg m_i$, and the energy $E_i \simeq p + \frac{m_i^2}{p}$, the oscillation probability reads

$$P_{\nu_e \to \nu_\mu}(t) \equiv |\langle \nu_\mu(o)|\nu_e(t)\rangle|^2 = \sin^2\theta \sin^2 \frac{\Delta m^2 L}{4E} . \tag{5.7}$$

Here $\Delta m^2 = |m_2^2 - m_1^2|$, E is the average energy of the neutrino beam, and L is the length traversed by the beam before a ν_μ detection occurs. The quantity $L^0 = \frac{4\pi E}{\Delta m^2}$ is the oscillation length and corresponds to the distance travelled by a neutrino to return to the initial state (i.e. the phase becomes 2π). In a similar way, the survival probabilities that a ν_e/ν_μ is detected as a ν_e/ν_μ beam after a time t are $P_{\nu_e \to \nu_e}(t) = P_{\nu_\mu \to \nu_\mu}(t) = 1 - P_{\nu_e \to \nu_\mu}(t)$.

The generalization to three flavours is straightforward. The mass eigenstates $|\nu_i\rangle$ evolves with time as $|\nu_i\rangle(t) = e^{-iE_it}|\nu_i\rangle(0)$, while the weak eigenstates evolve according to

$$|\nu_\alpha\rangle = \sum_i^3 U_{\alpha i} e^{-iE_it}|\nu_i\rangle(0) = \sum_i^3 U_{\alpha i} e^{-iE_it} U^*_{\beta i}|\nu_\beta\rangle(0) \,. \tag{5.8}$$

The oscillation probability that a ν_β is present in a neutrino beam of ν_α at the instant t is given by

$$P_{\nu_\alpha \to \nu_\beta} = |\langle \nu_\beta(0)|\nu_\alpha(t)\rangle|^2 = \sum_{ij} |U_{\alpha i} U^*_{\beta i} U^*_{\alpha j} U_{\beta j}| \cos\left[(E_i - E_j)t - \Phi_{\alpha\beta ij}\right] \,, \tag{5.9}$$

where $\Phi_{\alpha\beta ij} = arg[U_{\alpha i} U^*_{\beta i} U^*_{\alpha j} U_{\beta j}]$, while the oscillation length is given by $L^0_{ij} = \frac{4\pi E}{\Delta m^2_{ij}}$.

5.2 General Aspects of Neutrino Spin-Flavour and Spin-Flip

According to the SM, neutrinos are left polarized particles. If the action of an external background allows neutrinos to change their polarization, then they acquire a right-handed polarization hence become sterile, and cannot be observed, inducing a suppression of the flux of the emitted left-handed neutrinos. Such studies, involving both neutrino spin and spin-flavour oscillations in various external fields also provide information about neutrino magnetic moments [1,2]. The electro-weak interaction of neutrino with background matter influences the process of neutrino oscillations [3], while the interaction of neutrinos with gravitational fields generated by astrophysical objects, though weak, can also affect the dynamics of neutrino oscillations. In this connection, the influence of gravitational fields on neutrino spin oscillations has been investigated in [4–6](see also Ref. [7–10]).

These effects can take place in the dense and hot matter of the accretion disk around a supermassive black hole where, besides photons, significant fluxes of neutrinos can be present. As shown in [11], these neutrinos may affect the r-process nucleosynthesis near a black hole surrounded by an accretion disk.

The Standard Model of particle physics predicts that neutrinos do not possess a magnetic moment. However, if neutrinos possess a magnetic moment, then they undergo a spin-flip while passing through a magnetic field [12–16], as well as spin-precession and spin-flavour oscillations. Non-standard neutrino properties, such as mass and magnetic moment, are extremely important because they occur in Nature and also in many Grand Unified Theories [17].

The magnetic moment μ of the neutrinos is subject to several constraints. Terrestrial experiments give $\mu < 10^{-10}\mu_B$ [18–20], where μ_B is the Bohr magneton, while astrophysical observations provide more stringent bounds $\mu < 2 \times 10^{-12}\mu_B$ [21–25] (these values are, however, model-dependent).

It is a consolidated fact that neutrinos play a fundamental role in type II Supernovae due to the fact that they drive the Supernova explosion [26,27]. In this context neutrino oscillations are expected to affect the explosion mechanism directly [28]. In fact, the neutrino spectrum has allowed the testing of various oscillation scenarios [29,30]), and neutrino spin-flips provide a useful mechanism to explain the high pulsar kick velocities [31]. It is also well known that gyroscopes precess in strong gravitational fields [32,33].

Some of the conditions under which gravitation can influence the neutrino spin-flavour and spin behaviour are investigated in this chapter.

5.2.1 The Equation of Evolution in Curved Space–Time

As seen in the previous chapters, gravitational effects on spin arise through the spin connection Γ_μ entering the Dirac equation [34–37]. The covariant Dirac equation for massive neutrinos interacting minimally with the gravitational field [38] is given by

$$\left[\gamma^a e_a^\mu(\partial_\mu + \Gamma_\mu) + M\right]\psi = 0, \tag{5.10}$$

where two neutrino species are only considered for simplicity, ψ represents the column vector of spinors of different mass eigenstates, M is the mass matrix $M^2 = \text{diag}[m_1^2, m_2^2]$ and m_1 and m_2 are the mass eigenvalues of the two neutrino species. The explicit expression for Γ_μ is given by Eq. (1.53). Using Eq. (1.49), the non-vanishing contribution from the spin connection is again given by Eq. (4.129) and $A_{G\mu}$ by Eq. (4.130). In Eq. (4.129), left- and right-handed states are treated differently because of the presence of γ^5. Following [34], one can group these terms together with those arising from matter effects by adding a term proportional to the identity. In this way the spin connection term assumes the form $\gamma^a e_a^\mu \Gamma_\mu = \gamma^a e_a^\mu (i A_{G\mu} \mathcal{P}_L)$, where $\mathcal{P}_L = (1-\gamma^5)/2$. Notice that $A_{G\mu}$ is diagonal in spin space and does not therefore give rise to spin-flips, while it can change the resonance as well as the adiabaticity conditions, as shown below, in the presence of off-diagonal terms (for example, the inclusion of the interaction of particles with a magnetic field) [34]. The mass matrix M is related to the mass matrix in flavour space M_f through a rotation by the vacuum mixing angle θ_v according to

$$M_f^2 = U \begin{pmatrix} m_1^2 & 0 \\ 0 & m_2^2 \end{pmatrix} U^\dagger, \quad U = \begin{pmatrix} \cos\theta_v & \sin\theta_v \\ -\sin\theta_v & \cos\theta_v \end{pmatrix}. \tag{5.11}$$

Indicating with $\hat{H}$ the Hamiltonian in the chiral basis, the two neutrino flavour evolution equation can be written in the form

$$i\frac{d}{d\lambda}\begin{pmatrix} \nu_{eL} \\ \nu_{\mu L} \\ \nu_{eR} \\ \nu_{\mu R} \end{pmatrix} = \hat{H}\begin{pmatrix} \nu_{eL} \\ \nu_{\mu L} \\ \nu_{eR} \\ \nu_{\mu R} \end{pmatrix}, \tag{5.12}$$

$$\hat{H} = \begin{pmatrix} H_{ee} & H_{e\mu} \\ H_{\mu e} & H_{\mu\mu} \end{pmatrix}, \tag{5.13}$$

$$H_{ee} = \begin{pmatrix} V_c + V_n + \frac{P_\mu A^\mu_G}{E_l} - \frac{\delta m^2}{4E_l}\cos 2\theta_v & \frac{\delta m^2}{4E_l}\sin 2\theta_v \\ \frac{\delta m^2}{4E_l}\sin 2\theta_v & V_n + \frac{P_\mu A^\mu_G}{E_l} + \frac{\delta m^2}{4E_l}\cos 2\theta_v \end{pmatrix}, \tag{5.14}$$

$$H_{\mu\mu} = \begin{pmatrix} -\frac{\delta m^2}{4E_l}\cos 2\theta_v & \frac{\delta m^2}{4E_l}\sin 2\theta_v \\ \frac{\delta m^2}{4E_l}\sin 2\theta_v & \frac{\delta m^2}{4E_l}\cos 2\theta_v \end{pmatrix}, \tag{5.15}$$

$$H_{e\mu} = \begin{pmatrix} \mu_{ee} B & \mu_{e\mu} B \\ \mu_{\mu e} B & \mu_{\mu\mu} B \end{pmatrix}, \quad H_{\mu e} = \begin{pmatrix} \mu_{ee} B & \mu_{e\mu} B \\ \mu_{\mu e} B & \mu_{\mu\mu} B \end{pmatrix}, \tag{5.16}$$

where $\delta m^2 = m_2^2 - m_1^2$ and $\hat{H}$ includes the weak, electromagnetic and gravitational interactions. In other words, the interaction of neutrinos with matter, through the weak interaction $A^\mu_w \mathcal{P}_L$ written in the chiral basis $\nu_{eL}, \nu_{eR}, \nu_{\mu L}, \nu_{\mu R}$, and $A^\mu_w = \text{diag}(V_c + V_n, V_n, 0, 0)$ contains matter potentials due to the neutral and charged currents in neutral, unpolarized matter, given by

$$V_n = -\sqrt{2} G_F n_n \quad \text{and} \quad V_c = 2\sqrt{2} G_F n_e,$$

respectively. Here G_F is the Fermi constant, E_l the energy measured in a local inertial frame, and $n_{e,n}$ are the electron and neutron densities in the rest frame of background matter (in the above expression the ν_μ can be replaced by ν_τ). In (5.16) there appears the interaction of the neutrino with an external magnetic field $\hat{\mu}\bar{\psi}\sigma^{ab} F_{ab}\psi$, where $\hat{\mu}$ is the magnetic moment matrix and F_{ab} are the components of the electromagnetic field tensor (B is the component of the magnetic field which is perpendicular to the neutrino trajectory). The right-handed components are assumed to be sterile, so that they do not interact with matter.

Resonance occurs whenever the difference in diagonal elements vanishes. These can be computed from the expression for $\hat{H}$ (see Table 5.1) [35].

Table 5.1 Resonance conditions for spin-flips and MSW-conversions

Cases	Transitions	Resonance conditions
1	$\nu_{eL} \leftrightarrow \nu_{\mu L}$	$V_c - \frac{\delta m^2}{2E_l}\cos 2\theta_v = 0$
2	$\nu_{eL} \leftrightarrow \nu_{eR}$	$V_c + V_n + \frac{P_\mu A^\mu_G}{E_l} = 0$
3	$\nu_{eL} \leftrightarrow \nu_{\mu R}$	$V_c + V_n + \frac{P_\mu A^\mu_G}{E_l} - \frac{\delta m^2}{2E_l}\cos 2\theta_v = 0$
4	$\nu_{\mu L} \leftrightarrow \nu_{eR}$	$V_n + \frac{P_\mu A^\mu_G}{E_l} + \frac{\delta m^2}{2E_l}\cos 2\theta_v = 0$
5	$\nu_{\mu L} \leftrightarrow \nu_{\mu R}$	$V_n + \frac{P_\mu A^\mu_G}{E_l} = 0$

As seen from Table 5.1, Case 1, for example, corresponds to the active–active MSW transition, in which the matter potential is due to the charged current only and is proportional to the electron fraction $Y_e = n_e/(n_n + n_e)$, while, for the transitions corresponding to Cases 2 and 3, the potentials follow from the neutral as well as from the charged currents [39]. The *gravitational current* $P_\mu A^\mu_G$ takes into account the gravitational source. An explicit calculation is given for a static and for a rotating star.

5.2.2 Relativistic Effects Near a Static Star

The geometry in a spherically symmetric, static space–time is globally represented by the Schwarzschild line element

$$ds^2 = -e^{2\Phi(r)}dt^2 + e^{2\Lambda(r)}dr^2 + r^2 d\theta^2 + r^2 \sin^2\theta d\phi^2, \tag{5.17}$$

where $e^{2\Phi(r)} = e^{-2\Lambda(r)} = 1 - r_s/r$, $r_s = 2M$ is the Schwarzschild radius and M is the central gravitating mass. In this form, the tetrad is

$$e^\mu_a = \text{diag}\left[e^{-\Phi(r)}, e^{-\Lambda(r)}, \frac{1}{r}, \frac{1}{r\sin\theta}\right]. \tag{5.18}$$

The components of A^μ_G vanish identically, $A^\mu_G = 0$ [34], while the time-like component of the four-momentum $P_0 = -E$ is a conserved quantity. The energy measured by a locally inertial observer is gravitationally redshifted by the factor $1/\sqrt{1 - r_s/r}$, that is $E_l = Ee^{-\Phi(r)}$. MSW transitions in a Schwarzschild metric have been discussed in [34].

Consider, for example, the transition $\nu_{eL} \leftrightarrow \nu_{\mu R}$, Case 3 of the Table, and introduce the (helicity) mixing angle θ_m

$$\tan 2\theta_m = \frac{2\mu_{e\mu}B}{e^\Phi \delta m^2 \cos 2\theta_v/2E - (V_c + V_n)}, \tag{5.19}$$

which diagonalizes the corresponding submatrix in the Hamiltonian [17]. To quantify the magnitudes of the off-diagonal elements relative to those of the diagonal elements of $\hat{H}$ in the basis of the instantaneous eigenstates, one introduces the adiabaticity parameter γ corresponding to a particular transition. In order to have an adiabatic propagation, at the resonance point one must have $\gamma > 1$, where the explicit expression for $\gamma(r_{res})$ is

$$\gamma(r_{res}) = \frac{8e^{-\Phi(r_{res})}E\mu^2_{e\mu}B^2}{\delta m^2 \cos 2\theta_v e^{-\Lambda(r_{res})}|d\ln(V_c + V_n)/dr|_{res}}, \tag{5.20}$$

which exhibits the occurrence of the spatial dependence of the metric through the local energy. The conversion probability is given by [17]

$$P_{\nu_{eL}\to\nu_{\mu R}} = \frac{1}{2} - \left(\frac{1}{2} - P_{hop}\right)\cos 2\theta_v \cos 2\theta_m, \tag{5.21}$$

where P_{hop} is the non-adiabatic hopping probability, $P_{hop} = e^{-\frac{\pi}{2}\gamma(r_{\rm res})}$. The conversion probabilities corresponding to the other cases in Table 5.1 can be derived in a similar way.

5.2.3 Relativistic Effects Near a Rotating Star

If the gravitational source is rotating, the collapsing star may cause a dragging of the inertial frames and the line element is given by

$$ds^2 = -e^{2\Phi(r)}dt^2 + e^{2\Lambda(r)}dr^2 + r^2(d\theta^2 + \sin^2\theta d\phi^2) - 2\Omega r^2 \sin^2\theta d\phi dt, \tag{5.22}$$

to first order in the angular velocity Ω of the central mass (that is $\Omega a << 1$, where a is the radius of the central mass [40,41]). The off-diagonal terms can flip the spin of the neutrinos by transferring the angular momentum of the star to the neutrino spin. In the weak-field limit and for relativistic neutrinos moving radially outwards, for which $P_1 \approx P_0$, the gravitational term $P_\mu A_G^\mu / E_l$ assumes the form (using Eq. (5.22))

$$\frac{P_\mu A_G^\mu}{E_l} \approx e^{-\Lambda(r)} \frac{J}{r^3} \cos\theta \,. \tag{5.23}$$

Notice that for propagation in the equatorial plane, $\theta = \pi/2$, the gravitational current vanishes. Taking into account these results, Eqs. (5.20) and (5.21) become [42]

$$\tan 2\theta_m = \frac{2\mu_{e\mu} B}{e^{\Phi}\delta m^2 \cos 2\theta_v / 2E - (V_c + V_n + P_\mu A_G^\mu / E_l)}, \tag{5.24}$$

and

$$\gamma(r_{res}) = \frac{8e^{-\Phi(r_{res})} E \mu_{e\mu}^2 B^2}{\delta m^2 \cos 2\theta_v e^{-\Lambda(r_{res})} |d \ln(V_c + V_n + P_\mu A_G^\mu / E_l)/dr|_{res}}. \tag{5.25}$$

The term $P_\mu A_G^\mu$ changes sign at the equator, an effect related to an asymmetry in the neutrino emission. Moreover, such a term shifts the resonance point closer to the centre of the supernova in the upper hemisphere and further away in the lower hemisphere. A comment is in order. The analysis has been performed in the weak-field approximation and only radial neutrino trajectories have been considered. However, strong regimes arising from the space–time curvature can potentially become significant for spin-flip transitions in the supernova envelope. The extent to which they affect the neutrino transition depends on the detailed composition of the supernova, and on the mass and rotation rate of the proto-neutron star.

5.3 Spin Oscillations of Neutrinos Scattered Off a Rotating Black Hole

Consider now the spin oscillations of neutrinos using a semiclassical approach [4,5]. Consider, more specifically, the variation of the helicity, $h = \boldsymbol{\zeta} \cdot \mathbf{u}/|\mathbf{u}|$, where $\boldsymbol{\zeta}$ is the neutrino invariant spin and $\mathbf{u}$ is the spatial part of the neutrino four-velocity in the locally Minkowski frame, for neutrinos scattered gravitationally off a black hole. The aim is to derive the probabilities of spin oscillations for neutrinos interacting with Schwarzschild and Kerr gravitational backgrounds. Following Refs. [4,5], the evolution of $\boldsymbol{\zeta}$ in an external gravitational field is given by

$$\frac{\mathrm{d}\boldsymbol{\zeta}}{\mathrm{d}t} = 2(\boldsymbol{\zeta} \times \boldsymbol{\Omega}_g), \tag{5.26}$$

where $\boldsymbol{\zeta}$ is defined in a locally Minkowski frame, t is the time in world coordinates and $\boldsymbol{\Omega}_g$ accounts for the gravity contribution. As shown in [4,5], if a neutrino interacts with a Schwarzschild or a Kerr black hole, then the non-vanishing components of the vector $\boldsymbol{\Omega}_g$ in Eq. (5.26) are $\boldsymbol{\Omega}_g = (0, \Omega_2, 0)$. The non-vanishing neutrino spin components solution of (5.26) are ζ_1 and ζ_3 that can be conveniently parameterized in terms of the angle α between $\boldsymbol{\zeta}$ and the positive direction of the x-axis in the locally Minkowski frame,

$$\boldsymbol{\zeta} = (\cos\alpha, 0, \sin\alpha)\,. \tag{5.27}$$

The initial conditions are the following: at $r \to -\infty$, the incoming neutrino is left polarized $h_{-\infty} = \boldsymbol{\zeta}_{-\infty} \cdot \mathbf{u}_{-\infty}/|\mathbf{u}_{-\infty}| = -1$ (the helicity is negative), which implies $\boldsymbol{\zeta} = (-1, 0, 0)$, or $\alpha_{-\infty} = 0$. For the outgoing neutrino, at $r \to +\infty$, the helicity is given by $h_{+\infty} = \boldsymbol{\zeta}_{+\infty} \cdot \mathbf{u}_{+\infty}/|\mathbf{u}_{+\infty}|$, where $\boldsymbol{\zeta}_{+\infty} = (\cos\alpha_{+\infty}, 0, \sin\alpha_{+\infty})$. In these expressions, the asymptotic velocities $\mathbf{u}_{\pm\infty}$ are given by $\mathbf{u}_{\pm\infty} = \left(\pm\sqrt{E^2 - m^2}, 0, 0\right)$. From Eq. (5.27), one gets $h_{+\infty} = \cos\alpha_{+\infty}$, while the transition probability P_{LR} and survival probability P_{LL} for neutrino spin oscillations are given by

$$P_{\mathrm{LR}} = \frac{1}{2}(1 + \cos\alpha_{+\infty})\,, \quad P_{\mathrm{LL}} = \frac{1}{2}(1 - \cos\alpha_{+\infty})\,. \tag{5.28}$$

In the case of the Schwarzschild geometry (5.17), assuming that a neutrino moves in the equatorial plane with $\theta = \pi/2$, i.e. $d\theta = 0$, one finds that the non-vanishing component of $\boldsymbol{\Omega}_g$ is [4]

$$\Omega_2 = \frac{Le^{2\Phi}}{2Er^2}\left(-e^{\Phi} + \frac{U^t}{1 + U^t e^{\Phi}}\frac{r_g}{2r}\right), \tag{5.29}$$

where L is the conserved angular momentum of a neutrino, $U^t = dt/d\tau = Ee^{-2\Phi}/m$ is the component of the four-velocity U^μ in world coordinates, while the expression for $\mathbf{u}$ is [4]

$$\mathbf{u} = \left(\pm \frac{1}{m} \left[E^2 - m^2 e^{2\Phi} \left(1 + \frac{L^2}{m^2 r^2} \right) \right]^{1/2}, 0, \frac{L}{mr} \right). \tag{5.30}$$

The signs $\pm$ refer to outgoing and incoming neutrinos respectively. In the asymptotic limit, their corresponding velocities are $\mathbf{u}_{\pm\infty}$. The evolution of the angle α is obtained from (5.26) and (5.29) and is given by

$$\frac{d\alpha}{dr} = \pm \frac{L}{mr^2} \frac{\frac{E}{m}\left(\frac{3r_g}{2r} - 1\right) - 3e^{\Phi}}{\frac{E}{m} + e^{\Phi}} \left[\frac{E^2}{m^2} - e^{2\Phi} \left(1 + \frac{L^2}{m^2 r^2} \right) \right]^{-1/2}. \tag{5.31}$$

Introducing the quantities $x = r/r_g$, $y = b/r_g$, the impact parameter $b = L/E\sqrt{1 - \gamma^{-2}}$, and the Lorentz factor at infinity $\gamma = E/m$ one gets, for ultra-relativistic neutrinos $\gamma \gg 1$,

$$\alpha_{+\infty} = y \int_{x_m}^{\infty} \frac{dx(3 - 2x)}{x\sqrt{(x-1)R_S(x)}}, \tag{5.32}$$

where x_m is the minimum of the equation $R_S(x) = 0$, with $R_S(x) = x^3 - y^2(x - 1)$. The general expression for α_∞ can be found in [43]. As shown in [44], by evaluating $dR_S(x)/dx = 0$ and setting $R_S(x) = 0$, one finds that $3\sqrt{3}/2 \leq y < \infty$. For these values one gets $\alpha_{+\infty} = -\pi$, which implies, using Eq. (5.28), that $P_{LR} = (1 + \cos\alpha_{+\infty})/2 = 0$. Therefore, there is no spin-flip of ultra-relativistic neutrinos when they scatter off a Schwarzschild black hole (see also [45]).

Consider now the spin evolution of neutrinos scattered off a rotating black hole described by the Kerr metric, that in Boyer–Lindquist coordinates (t, r, θ, ϕ) is [46],

$$d\tau^2 = \left(1 - \frac{rr_g}{\Sigma}\right) dt^2 + 2\frac{rr_g a \sin^2\theta}{\Sigma} dt d\phi - \frac{\Sigma}{\Delta} dr^2 - \Sigma d\theta^2 - \frac{\Xi}{\Sigma} \sin^2\theta d\phi^2, \tag{5.33}$$

where

$$\Delta = r(r - r_g) + a^2, \quad \Sigma = r^2 + a^2 \cos^2\theta, \quad \Xi = \left(r^2 + a^2\right)\Sigma + rr_g a^2 \sin^2\theta. \tag{5.34}$$

The parameter a varies in the range $0 < a < r_g/2$. The black hole angular momentum is $J = Ma$. Moreover, the neutrinos move in the equatorial plane of a Kerr black hole with $\theta = \pi/2$ and $d\theta = 0$. The energy E and the neutrino angular momentum L are still integrals of motion. Repeating the calculation as in the case of the Schwarzschild geometry, in the case of ultra-reletivistic particles $\gamma \gg 1$, one finds

$$\alpha_{+\infty} = \int_{x_m}^{\infty} F_K(x) dx, \tag{5.35}$$

where

$$F_{\rm K}(x) \equiv y \frac{x^4(3-2x)+xyz(3-4x)-xz^2(3-3x+2x^2)-2yz^3+2z^4}{\sqrt{R(x)[x(x-1)+z^2][x^3+z^2(x+1)][x^3+z^2(x+1)-zy]}}, \tag{5.36}$$

with $z = a/r_g$ and

$$R_{\rm K}(x) \to x^3 + (z^2 - y^2)x + (y-z)^2. \tag{5.37}$$

The limit $z = 0$ reproduces the results derived for the Schwarzschild geometry.

As discussed in [44], the most general case in which neutrinos interact (in the locally Minkowskian frame) with gravitational and electromagnetic fields and background matter (that is neutrinos scattering off a rotating black hole with accretion disk permeated by a magnetic field), is described by the following generalization of the spin-precession Eq. (5.26)

$$\frac{{\rm d}s^a}{{\rm d}t} = \frac{1}{U^t}\left[G^{ab}s_b + 2\mu\left(f^{ab}s_b - u^a u_b f^{bc}s_c\right) + \sqrt{2}G_{\rm F}\varepsilon^{abcd}g_b u_c s_d\right], \tag{5.38}$$

where $G_{ab} = \gamma_{abc}u^c$ is the antisymmetric tensor that accounts for the gravitational interaction of neutrinos, $\gamma_{abc} = \eta_{ad}e^d_{\ \mu;\nu}e_b^{\ \mu}e_c^{\ \nu}$ are the Ricci rotation coefficients expressed in terms of the vierbein fields associated to the metric (5.33)

$$e_0^{\ \mu} = \left(\sqrt{\frac{\Xi}{\Sigma\Delta}}, 0, 0, \frac{arr_g}{\sqrt{\Delta\Sigma\Xi}}\right), \quad e_1^{\ \mu} = \left(0, \sqrt{\frac{\Delta}{\Sigma}}, 0, 0\right),$$
$$e_2^{\ \mu} = \left(0, 0, \frac{1}{\sqrt{\Sigma}}, 0\right), \quad e_3^{\ \mu} = \left(0, 0, 0, \frac{1}{\sin\theta}\sqrt{\frac{\Sigma}{\Xi}}\right), \tag{5.39}$$

$g^a = e^a_{\ \mu}G^\mu = (g^0, \mathbf{g})$ is the effective potential of the neutrino matter interaction in the locally Minkowski frame, $f_{ab} = e_a^{\ \mu}e_b^{\ \nu}F_{\mu\nu} = (\mathbf{e}, \mathbf{b})$ is the electromagnetic field tensor in the locally Minkowski frame, and $G_{\rm F} = 1.17 \times 10^{-5}\,{\rm GeV}^{-2}$ is the Fermi constant. The spin four vector s^a is related to the invariant three vector $\boldsymbol{\zeta}$ of the neutrino spin, which describes the polarization in the neutrino rest frame, by the relation

$$s^a = \left((\boldsymbol{\zeta}\cdot\mathbf{u}), \boldsymbol{\zeta} + \frac{\mathbf{u}(\boldsymbol{\zeta}\cdot\mathbf{u})}{1+u^0}\right). \tag{5.40}$$

The spin-precession is

$$\frac{d\boldsymbol{\zeta}}{{\rm d}t} = 2(\boldsymbol{\zeta}\times\boldsymbol{\Omega}), \tag{5.41}$$

where $\mathbf{\Omega}$ is given by

$$\mathbf{\Omega} = \frac{1}{U^t}\left\{\frac{1}{2}\left[\mathbf{b}_g + \frac{1}{1+u^0}(\mathbf{e}_g \times \mathbf{u})\right] + \frac{G_F}{\sqrt{2}}\left[\mathbf{u}\left(g^0 - \frac{(\mathbf{gu})}{1+u^0}\right) - \mathbf{g}\right] + \mu\left[u^0\mathbf{b} - \frac{\mathbf{u}(\mathbf{ub})}{1+u^0} + (\mathbf{e}\times\mathbf{u})\right]\right\}. \quad (5.42)$$

The quantity $\mathbf{e}_g$ and $\mathbf{b}_g$ are the components of the tensor G_{ab}, that is $G_{ab} = (\mathbf{e}_g, \mathbf{b}_g)$.

For purely gravitational neutrino scattering one gets $\alpha_{+\infty}^{(g)} = -\pi$ for any $z \leq 1/2$ and $y > y_0 = 4\cos^3\left[\frac{1}{3}\arccos(\mp 2z)\right] \pm z$. This means that $P_{\mathrm{LR}} = 0$, i.e. there are no spin oscillations of ultra-relativistic neutrinos. Ultra-relativistic neutrinos may however precess when magnetic fields are present.

5.4 Lensing and Oscillations Probability

In all processes in which the fundamental properties of neutrinos are studied, it is implied that the neutrino oscillation probabilities depend only on the difference of the square of neutrino masses, but not on their individual masses, that is the oscillation experiments do not allow the determination of the absolute neutrino mass. Moreover, in the current experiments one may measure the mass-squared difference Δm_{21}^2 and $|\Delta m_{31}^2|$, where $\Delta m_{ij}^2 \equiv m_i^2 - m_j^2$, $i, j = 1, 2, 3$, which offers two possibilities: the normal hierarchy $m_1 < m_2 < m_3$ and the inverse one $m_3 < m_1 < m_2$.

The possibility is explored here that the gravitational lensing of neutrinos may reveal the absolute scale of neutrino masses [47]. This is based on the observation that neutrinos travel from the source S to the observation point (detector D) following different trajectories around a gravitating source. That is, they get lensed at a common point and arrive at the detector following different path lengths, which results in interference of the oscillation probabilities at the point of observation. The oscillations then depend on the difference of the squared masses and on the absolute masses of the neutrinos.

5.4.1 Gravitational Lensing of Neutrinos in Schwarzschild Space–Time

Consider a black hole as a gravitational lens situated between the neutrino S and D. The geometry is described by the Schwarzschild metric. The neutrino flavour eigenstate ν_α in a curved space–time evolves as

$$|\nu_\alpha(t_D, \mathbf{x}_D)\rangle = N \sum_i U_{\alpha i}^* \sum_p e^{-i\Phi_i^p} |\nu_i(t_S, \mathbf{x}_S)\rangle\,, \quad (5.43)$$

where $|\nu_i(t_S, \mathbf{x}_S)\rangle$ is the neutrino mass eigenstate, $\sum_p$ indicates the sum over all trajectories p, and Φ_j^p denotes the phase given by (assuming the weak-field approximation, $GM \ll r$ and $b \ll r_{S,D}$, see Eq. (C.11)) [48]

$$\Phi_j^p \simeq \frac{m_j^2}{2E_0}(r_S + r_D)\left(1 - \frac{b_p^2}{2r_S r_D} + \frac{2GM}{r_S + r_D}\right), \tag{5.44}$$

where b_p is the impact parameter corresponding to the p-trajectory. The probability of flavour transitions is [47]

$$P_{\alpha\beta}^{\text{lens}} = |\langle\nu_\beta|\nu_\alpha(t_D, \mathbf{x}_D)\rangle|^2 = |N|^2 \sum_{i,j} U_{\beta i} U_{\beta j}^* U_{\alpha j} U_{\alpha i}^* \sum_{p,q} e^{-i\Delta\Phi_{ij}^{pq}}, \tag{5.45}$$

where the normalization factor is defined as

$$|N|^2 = \frac{1}{\sum_i |U_{\alpha i}|^2 \sum_{p,q} \exp\left(-i\Delta\Phi_{ii}^{pq}\right)}. \tag{5.46}$$

From (5.44) one can write the phase difference in two parts, depending on the mass-squared difference Δm_{ij}^2 and on the path difference Δb_{pq}^2, that is

$$\Delta\Phi_{ij}^{pq} = \Phi_i^p - \Phi_j^q = \Delta m_{ij}^2 A_{pq} + \Delta b_{pq}^2 B_{ij}, \tag{5.47}$$

where

$$A_{pq} = \frac{(r_S + r_D)}{2E_0}\left(1 + \frac{2GM}{r_S + r_D} - \frac{\sum b_{pq}^2}{4r_S r_D}\right),$$
$$B_{ij} = -\frac{\sum m_{ij}^2}{8E_0}\left(\frac{1}{r_S} + \frac{1}{r_D}\right). \tag{5.48}$$

Here, $\sum b_{pq}^2 = b_p^2 + b_q^2$ and $\sum m_{ij}^2 = m_i^2 + m_j^2$. The oscillation probability is given by

$$P_{\alpha\beta}^{\text{lens}} = \frac{\sum_{i,j} U_{\beta i} U_{\beta j}^* U_{\alpha j} U_{\alpha i}^* \left[\sum_q e^{-i\Delta m_{ij}^2 A_{qq}} + \sum_{q>p} 2\cos\left(\Delta m_{ij}^2 A_{pq} + \Delta b_{pq}^2 B_{ij}\right)\right]}{N_{\text{path}} + \sum_i |U_{\alpha i}|^2 \sum_{q>p} 2\cos(\Delta b_{pq}^2 B_{ii})} \tag{5.49}$$

with $\sum_\beta \mathcal{P}_{\alpha\beta}^{\text{lens}} = 1$. Assuming for simplicity, that neutrinos propagate in the plane $\theta = \pi/2$ plane, and considering that $N_{\text{path}} = 2$ in the simplest case of two neutrino flavours, then the probability for the $\nu_e \to \nu_\mu$ transition obtained from (5.49) is

$$\begin{aligned} P_{e\mu}^{\text{lens}} = |N|^2 \sin^2 2\alpha \Big[& \sin^2\left(\Delta m^2 \frac{A_{11}}{2}\right) + \sin^2\left(\Delta m^2 \frac{A_{22}}{2}\right) \\ & - \cos(\Delta b^2 B_{12})\cos(\Delta m^2 A_{12}) + \frac{1}{2}\cos(\Delta b^2 B_{11}) + \frac{1}{2}\cos(\Delta b^2 B_{22})\Big], \end{aligned} \tag{5.50}$$

where

$$|N|^2 = \frac{1}{2\left(1+\cos^2\alpha\cos(\Delta b^2 B_{11})+\sin^2\alpha\cos(\Delta b^2 B_{22})\right)}, \tag{5.51}$$

and $\Delta m^2 = \Delta m^2_{21}$. The parameters $A_{11,22}$ and $B_{11,22}$ entering (5.50) are given by

$$A_{11,22} = A_{12} \mp X\Delta b^2\,, \quad B_{11,22} = B_{12} \mp X\Delta m^2\,,$$

where the sign $-$ refers to the 11-component and the sign $+$ to the 22-component, while

$$X = \frac{r_S + r_D}{8E_0 r_S r_D}\,.$$

The new features introduced by the lensing probability (5.50) [47] are: (1) The probability (5.50) is not invariant under the interchange $m_1 \leftrightarrow m_2$, unless $\Delta b^2 = 0$ or $\alpha = \pi/4$ (this is due to the $B_{11,22}$ terms). In flat space–time the probability is instead invariant under $m_1 \leftrightarrow m_2$. The consequence of this is that the oscillation probability is sensitive to the neutrino mass ordering, so that one obtains different probabilitics for $\Delta m^2 > 0$ and $\Delta m^2 < 0$. (2) In general, the lensing probability $\mathcal{P}^{\text{lens}}_{e\mu}$ depends explicitly on the sum of the squared neutrino masses $\sum m^2 = m_1^2 + m_2^2$ through the B_{12}-term. These features can be evidenced by means of the dimensionless parameter $\epsilon \equiv \Delta b^2 B_{12}$, with $\epsilon \ll 1$, by expanding (5.50) as

$$P^{\text{lens}}_{e\mu} \approx \sin^2 2\alpha \sin^2\left(\Delta m^2\frac{A_{12}}{2}\right)\left[1 - \frac{\epsilon^2}{2}\frac{\Delta m^2}{\sum m^2}\cos 2\alpha + \frac{\epsilon^4}{16}\frac{\Delta m^2}{\sum m^2}\right. \tag{5.52}$$
$$\left. \times\left(\frac{\Delta m^2}{\sum m^2}\left(2\cos 4\alpha + \csc^2\frac{\Delta m^2 A_{12}}{2}\right) - \frac{2}{3}\left(1+\left(\frac{\Delta m^2}{\sum m^2}\right)^2\right)\cos 2\alpha\right) + O(\epsilon^6)\right].$$

The probability of transition depends on $\sum m^2$, as well as on the sign of Δm^2, provided the mixing angle is not maximal, $\alpha \neq \pi/4$. If the mixing angle is maximal ($\alpha = \pi/4$), the probability (5.52) does not depend on the $\sum m^2$ at the second order in ϵ^2. In the latter case the mass dependent lensing effects arise in higher order terms in ϵ [47,48].

References

1. Fujikawa, K., Shrock, R.: Phys. Rev. Lett. **45**, 963 (1980)
2. Giunti, C., Kouzakov, K.A., Li, Y.-F., Lokhov, A.V., Studenikin, A.I., Zhou, S.: Ann. Phys. (Amsterdam) **528**, 198 (2016)
3. Bilenky, S.: Introduction to the Physics of Massive and Mixed Neutrinos, 2nd edn. Springer, Cham (2018)
4. Dvornikov, M.: Int. J. Mod. Phys. D **15**, 1017–1034 (2006)
5. Dvornikov, M.: J. Cosmol. Astropart. Phys. **06**, 015 (2013)
6. Dvornikov, M.: Phys. Rev. D **99**, 116021 (2019)
7. Obukhov, Yu.N., Silenko, A.J., Teryaev, O.V.: Phys. Rev. D **96**, 105005 (2017)

8. H.J. Mosquera Cuesta, G. Lambiase, Astrophys.J. 689 (2008) 371. H.J. Mosquera Cuesta, G. Lambiase Int. J. Mod. Phys. D **18**, 435 (2009)
9. Lambiase, G.: Mon. Not. Roy. Astron. Soc. **62**, 867 (2005)
10. H.J. Mosquera Cuesta, G. Lambiase, J.P. Pereira, Phys. Rev. D **95**, 025011 (2017)
11. Wanajo, S., Janka, H.-Th.: Astrophys. J. **746**, 180 (2012)
12. Lim, C.S., Marciano, W.J.: Phys. Rev. D **37**, 1368 (1988)
13. Voloshin, M.B.: Phys. Lett. B **209**, 360 (1988)
14. Akhmedov, EKh.: Phys. Lett. B **213**, 64 (1988)
15. J. Schechter and J.W.F. Valle, Phys. Rev. D **24**, 1883 (1981); erratum: Phys. Rev. D **25**, 283 (1982)
16. M.B. Voloshin, M.I. Vysotsky and L.B. Okun, Zh. Eksp. Teor. Fiz. **91**, 754 (1986) [Sov. Phys. JETP **64**, 446 (1986)]
17. Haxton, W.C.: Ann. Rev. Astron. Astrophys. **33**, 459 (1995)
18. Particle Data Group, Barnett, R.M., et al.: Phys. Rev. D **54**, 1658 (1996)
19. Vidyakin, G.S., et al.: JETP Lett. **55**, 206 (1992)
20. Vogel, P., Engel, J.: Phys. Rev. D **39**, 3378 (1989)
21. Nötzold, D.: Phys. Rev. D **38**, 1658 (1988)
22. Peltoniemi, J.T.: Astron. Astrophys. **254**, 121 (1992)
23. Lattimer, J.M., Cooperstein, J.: Phys. Rev. Lett. **61**, 23 (1988)
24. Raffelt, G.G., Weiss, A.: Astron. Astrophys. **264**, 536 (1992)
25. Raffelt, G.G.: Phys. Rev. Lett. **64**, 2856 (1990)
26. Bethe, H.A., Wilson, J.R.: Astrophys. J. **295**, 14 (1985)
27. Burrows, A.S.: Supernovae, edited by A.G. Petschek. Springer, New York (1990)
28. K.S. Hirata *et al.*, Phys. Rev. Lett. **58**, 1490 (1987); R.M. Bionta *et al.*, Phys. Rev. Lett. **58**, 1494 (1987)
29. Loredo, T.J., Lamb, D.Q.: Proceedings of the 14th Texas Symposium on Relativistic Astrophysics, Dallas, Texas, 1988, edited by E. Fenyves [Ann. (N.Y.) Acad. Sci. 571, (1989)]
30. Smirnov, AYu., Spergel, D.N., Bahcall, J.N.: Phys. Rev. D **49**, 1389 (1994)
31. Akhmedov, EKh., Lanza, A., Sciama, D.W.: Phys. Rev. D **56**, 6117 (1997)
32. Thirring, H.: Phys. Zs. **19**, 33 (1918)
33. Brill, D., Cohen, J.: Phys. Rev. **143**, 1011 (1966)
34. Cardall, C.Y., Fuller, G.M.: Phys. Rev. D **55**, 7960 (1996)
35. Píriz, D., Roy, M., Wudka, J.: Phys. Rev. D **54**, 1587 (1996)
36. D. V. Ahluwalia and C. Burgard, LA-UR-96-862, gr-qc/9603008 (1996); D.V. Ahluwalia and C. Burgard, LA-UR-96-2031, gr-qc/9606031 (1996)
37. Y. Grossman and H.J. Lipkin, WIS-96/27/Jun-PH, TAUP 2346-96, hep-ph/9607201 (1996)
38. See e.g. S.W. Weinberg, Gravitation and Cosmology. Wiley, New York (1972), Sec. 12.5; D.R. Brill and J.A. Wheeler, Rev. Mod. Phys. **29**, 465
39. Athar, H., Peltoniemi, J.T., Smirnov, AYu.: Phys. Rev. D **51**, 6647 (1995)
40. Hartle, J.B.: Astrophys. J. **150**, 1005 (1967)
41. C.W. Misner, K.S. Thorne, and J.A. Wheeler Gravitation, p. 699. Freeman, New York (1973)
42. Bruggen, M.: Phys. Rev. D **58**, 083002 (1998)
43. Dvornikov, M.: Phys. Rev. D **101**, 056018 (2020)
44. Dvornikov, M.: J. Cosmol. Astropart. Phys. **04**, 005 (2021)
45. Dolan, S., Doran, C., Lasenby, A.: Phys. Rev. D **74**, 064005 (2006)
46. Rezzolla, L.: An introduction to astrophysical black holes and their dynamical production. In: Haardt, F., Gorini, V., Moschella, U., Treves, A., Colpi, M. (eds.) Astrophysical Black Holes, pp. 24–29. Springer, Cham (2016)
47. Swami, H., Lochan, K., Patel, K.M.: Phys. Rev. D **102**, 024043 (2020)
48. Fornengo, N., Giunti, C., Kim, C.W., Song, J.: Phys. Rev. D **56**, 1895 (1997)

6 Radiative Processes, Spin Currents, Vortices

Abstract

This chapter deals with developments that have their origin in subjects introduced in Chaps. 1 and 2. Gravitational Berry phases enable the formulation of an EFA which alters the dispersion relations of the particles interacting with the gravitational field. Hence, processes normally forbidden for kinematical reasons become physical in the presence of a gravitational field, as shown in Sect. 6.1. Section 6.2 deals with spin currents. These are currents in which the flow of spin can be separated from that of charge. This possibility has stimulated interest in fundamental spin physics. Spin control is an important issue in spin-based electronics and in all phenomena based on spin transport and dynamics. EFA makes possible the investigation of the transfer of angular momentum between external non-inertial fields and electron spins by means of the spin current tensor. The wave functions considered in Chap. 1 contain singularities that give rise to vortices. These are of interest in astrophysics, dark matter physics, and laboratory experiments. They are considered in Sect. 6.3. Some aspects of Zitterbewegung, or trembling motion, are studied in Sect. 6.4. In 1930, Schroedinger found that the electron velocity is not a constant of motion. Such an effect in vacuo must be quantum in nature. The expected trembling frequency and amplitude are not attainable with present experimental means. Zitterbewegung is a subject of intense interest in solid-state physics where its presence has been studied in crystalline solids, semiconductors, and graphene [1].

6.1 Radiative Processes

Kinematically forbidden processes may be allowed in the presence of external gravitational fields. These can be taken into account by introducing generalized particle momenta. The corresponding transition probabilities can then be calculated to all orders in the metric deviation from the field-free expressions by simply replacing the particle momenta with their generalized counterparts. The procedure applies to

G. Lambiase and G. Papini, *The Interaction of Spin with Gravity in Particle Physics*, Lecture Notes in Physics 993, https://doi.org/10.1007/978-3-030-84771-5_6

particles of any spin and to any gravitational fields. Transition probabilities, emission power, and spectra are, to leading order, linear in the metric deviation. It is also shown how a small dissipation term in the particle wave equations can trigger a strong back-reaction that introduces resonances in the radiative process and deeply affects the resulting gravitational background.

Processes in which massive, on-shell particles emit a photon according to Fig. 6.1 are examples of kinematically forbidden transitions that remain so unless the dispersion relations of at least one of the particles involved are altered. This possibility presents itself when particles travel in a medium or in an external gravitational field.

The action of gravitational fields on a particle's dispersion relations can be studied by solving the respective covariant wave equation. As shown in Chap. 1, this can be done exactly to first order in the metric deviation [2–7]. The solutions contain the gravitational field in a phase operator that alters in effect a particle four momentum by acting on the wave function of the field-free equations. This result applies equally well to fermions and bosons and can be extended to all orders in $\gamma_{\mu\nu}$. The calculation of even the most elementary Feynman diagrams does therefore require an appropriate treatment when external gravitational fields are present. While the inclusion of external electromagnetic fields has met with success in the case of static (Coulomb) fields [8], and can be easily carried out for static (Newtonian) fields, no systematic attempts have been made for relativistic gravity. The procedure developed below is intended to fill in part this gap and applies to weak, time-independent and time-dependent gravitational fields.

Assume, for simplicity, that P in Fig. 6.1 is an incoming fermion and that the photon ℓ and outgoing fermion p' are produced on-shell.

It is convenient to rewrite the solution of the covariant Dirac equation in the form

$$\Psi(x) = g(x)e^{-ipx}\, u_0(\mathbf{p})\,, \tag{6.1}$$

where

$$g(x) = \frac{1}{2m}\left[\left(\gamma^{\mu}(p_{\mu} + h_{\mu}^{\hat{\alpha}}(x)p_{\hat{\alpha}} + \Phi_{G,\mu}(x)\right) + m\right] e^{-i\Phi_T}\,, \tag{6.2}$$

and

$$\Phi_{G,\mu} = -\frac{1}{2}\int_P^x dz^{\lambda}(\gamma_{\mu\lambda,\beta} - \gamma_{\beta\lambda,\mu})p^{\beta} + \frac{1}{2}\gamma_{\alpha\mu}p^{\alpha}\,. \tag{6.3}$$

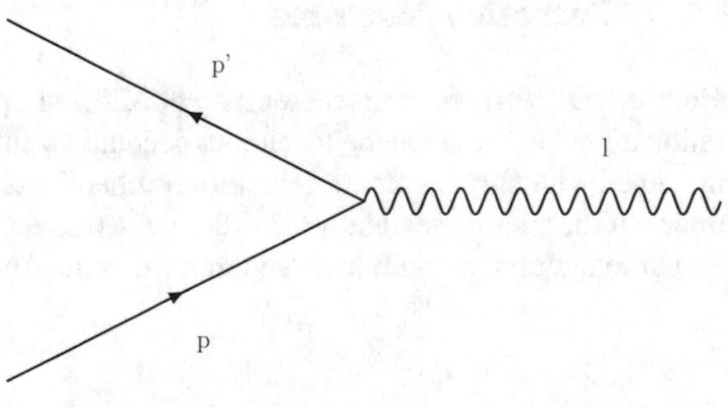

Fig. 6.1 p' and ℓ are the outgoing fermion and photon and p indicates the incoming fermion

It also is convenient to focus on the simple process of Fig. 6.1. The transition amplitude can be calculated by introducing the generalized four momentum

$$P_\mu = p_\mu + \tilde{h}_\mu^{\hat{\alpha}} p_{\hat{\alpha}} + \tilde{\Phi}_{G,\mu} - \frac{i}{2}(\tilde{\Phi}_G - \tilde{\Phi}_G^*) p_\mu \equiv p_\mu + \tilde{P}_\mu \,, \qquad (6.4)$$

for the incoming fermion, as (6.2) itself suggests. The part that contains the gravitational field is indicated by $\tilde{P}_\mu$. In (6.4), $\tilde{h}_\mu^{\hat{\alpha}}$, $\tilde{\Phi}_{G,\mu}$, and $\tilde{\Phi}_G$ are quantities that must be calculated, once the metric is known. They are related to the Fourier transforms of the corresponding expressions that appear in (1.71) and (6.3). P_μ is not on-shell. In fact

$$P^\mu P_\mu \equiv m_e^2 = m^2 + 2\left[p^\mu h_\mu^{\hat{\alpha}} p_\alpha + p^\mu \Phi_{G,\mu} - \frac{im^2}{2}(\Phi_G - \Phi_G^*) \right] , \qquad (6.5)$$

where $p_\mu p^\mu = m^2$ because p_μ is the momentum of the free fermion represented by $\Psi_0(x)$. The transition amplitude is then

$$M_1 = -iZe\eta^{\mu\nu} \bar{u}_0(\mathbf{p}') \varepsilon_{\hat{\mu}(\lambda)} \gamma_{\hat{\nu}} g(|\mathbf{q}|) u_0(\mathbf{p}) \,, \qquad (6.6)$$

where $\mathbf{q} \equiv \mathbf{p} - \mathbf{p}' - \boldsymbol{\ell}$, $\varepsilon^{\hat{\mu}}_{(\lambda)}$ represents the polarization of the photon, and Ze is the charge of the fermion. A transition amplitude M_2 must be added to M_1 to account for the fact that the contraction in (6.6) is, in general, accomplished by means of $g^{\mu\nu}$. It has been repeatedly calculated in the literature and is given by [9]

$$M_2 = -iZe\gamma^{\mu\nu}(|\mathbf{q}|) \bar{u}_0(\mathbf{p}') \varepsilon_{\hat{\mu}(\lambda)} \gamma_{\hat{\nu}} u_0(\mathbf{p}) \,. \qquad (6.7)$$

M_1 contains the part p_μ of (6.4) that comes from Ψ_0, and a new part that contains the gravitational contribution due to the propagation of the fermion in the field of the source. The total transition amplitude is given by $M = M_1 + M_2$.

The calculation now requires that a metric be selected. Consider the particular instance of a fermion that is propagating with momentum $p^3 \equiv p$, impact parameter $b \geq R$ and $x_2 = 0$, from $x_3 = -\infty$ towards a gravitational source of mass M and radius R placed at the origin and described by the lensing metric [7,10]

$$\gamma_{00} = 2\phi \,, \quad \gamma_{ij} = 2\phi\delta_{ij} \,,$$

where $\phi = -\frac{GM}{r}$. One finds

$$\Gamma_0 = -\frac{1}{2}\phi_{,j}\sigma^{0j} \,, \quad \Gamma_i = -\frac{1}{2}\phi_{,j}\sigma^{ij} \,,$$

and

$$e_{\hat{i}}^0 = 0 \,, \quad e_{\hat{0}}^0 = 1 - \phi \,, \quad e_{\hat{k}}^l = (1 + \phi)\,\delta_k^l \,.$$

All spin matrices are now expressed in terms of ordinary, constant Dirac matrices. It is also assumed that the on-shell conditions $p'_\mu p'^\mu = m^2$, $\ell_\mu \ell^\mu = 0$ remain valid. Extension of the calculation to include different particles or higher order gravitational contributions to p' , ℓ can be derived to all orders in $\gamma_{\mu\nu}$.

The Fourier transforms of the quantities that appear in (6.2) must now be calculated. One gets

$$
\begin{aligned}
h_0^{\hat{\alpha}}(q)p_\alpha &= 8\pi^2\delta(q_0)\delta(q_x)\delta(q_y)p_0 GMK_0(bq_z)\,,\, h_3^{\hat{\alpha}}(q)p_\alpha = \\
&= 8\pi^2\delta(q_0)\delta(q_x)\delta(q_y)pGMK_0(bq_z)\,,
\end{aligned} \tag{6.8}
$$

$$
\begin{aligned}
\Phi_{G,1}(q) + \Phi^*_{G,1}(q) &= 0\,, \\
\Phi_{G,2}(q) &= 0\,, \\
\Phi_{G,3}(q) &= -8\pi^2\delta(q_0)\delta(q_x)\delta(q_y)\left(\frac{p_0^2}{p} + p\right) GMK_0(bq_z)\,,
\end{aligned} \tag{6.9}
$$

and

$$
\Phi_G(q) = -i\,16\pi^2\delta(q_0)\delta(q_x)\delta(q_y)\left(\frac{2p_0^2}{p} - p_0 + p\right)\frac{GM}{bq_z}K_0(bq_z)\,. \tag{6.10}
$$

Four-momentum conservation to zeroth order only is required because (6.8), (6.9) and (6.10) are already of $O(\gamma_{\mu\nu})$. The Bessel function

$$
K_0(bq_z) \simeq \sqrt{\pi/2bq_z}e^{-bq_z}\left[1 - \frac{1}{8bq_z} + ...\right],
$$

itself a distribution, can be approximated by $K_0(bq_z) \simeq \delta(bq_z)$ and $\delta^4(q)$ from (6.8), (6.9), and (6.10) can be eliminated. Conservation of energy momentum will reappear as a factor $(2\pi)^4\delta^4(q)$ in the expression for the radiated power W defined below. The term $\Phi_G(q)$ does however diverge even more rapidly than the other Fourier transforms for small momentum transfers and will be dropped in what follows. This behaviour is well known and is related to the infinite range of the Newtonian potential and the use of a plane wave for $\Psi_0(x)$. By removing $\delta^4(q)$ from $h_\mu^{\hat{\alpha}}$ and $\Phi_{G,\mu}(q)$ one obtains $\tilde{h}_\mu^{\hat{\alpha}}$ and $\tilde{\Phi}_{G,\mu}$ of (6.4). One also finds

$$
\begin{aligned}
P_0 &\simeq p_0 + 4\pi\frac{GM}{b}p_0 = p_0 + \tilde{P}_0\,,\; P_1 \simeq 4\pi\frac{GM}{b}\left(\frac{p_0^2}{p} + p\right) = \tilde{P}_1\,, \\
P_2 &= 0\,, \qquad P^3 \equiv P = p - 4\pi\frac{GM}{b}p = p + \tilde{P}\,.
\end{aligned} \tag{6.11}
$$

The power radiated as photons in the process of Fig. 6.1 is calculated according to the formula [11]

$$W = \frac{1}{8(2\pi)^2} \int \delta^4(P - p' - \ell) \frac{|M|^2}{P p'_0} d^3 p' d^3 \ell \,. \tag{6.12}$$

There are two ways to calculate $|M_1|^2$. In the first one, p_α can be replaced with P_α in the field-free ($\gamma_{\mu\nu} = 0$) expression given by

$$\Sigma |M_1|^2 = Z^2 e^2 [-4m^2 (p'_\alpha p^\alpha) + 8(p_\alpha p^\alpha)] \,.$$

The gravitational contribution to M_1 then appears in $\tilde{P}_\mu$ exclusively. The terms $-32m^2(p'_\alpha p^\alpha) + 64m^2$ that do not contain gravitational contributions and therefore refer to the kinematically forbidden transition must also be removed. This yields, to $O(\gamma_{\mu\nu})$, the expression

$$\Sigma |M_1|^2 = Z^2 e^2 \left[-4(p'_\alpha \tilde{P}^\alpha) + 8(p_\alpha \tilde{P}^\alpha) \right] \,. \tag{6.13}$$

In a second, alternate approach, $|M_1|^2$ is calculated directly ($\gamma_{\mu\nu} \neq 0$) from (6.6). By summing over final spins and averaging over initial spins and polarizations, one obtains

$$\Sigma |M_1|^2 = \frac{Z^2 e^2}{2(2m)^2} Tr X_1 \tag{6.14}$$

$$X_1 \equiv (\not{p}' + m)\gamma_\beta \left[(\not{p} + \tilde{\not{P}} + m) \left((\not{p} + m)^2 + (\not{p} + m)\tilde{\not{P}}^* + H(\not{p} + m) \right) \right] \gamma^\beta \,,$$

where $\not{p} = \gamma^\mu p_\mu$ and

$$H = \frac{p^0}{p^3} \phi \left(\gamma^3 p^0 - \gamma^0 p^3 \right) + \left(\frac{p^0}{p^3} \gamma^1 p^0 - \gamma^1 p^3 \right) 4GM K_1(bq_z) \,.$$

On carrying out the traces of the Dirac matrices, the contribution from H vanishes. A similar, but simpler calculation, gives $|M_2|^2$. By further eliminating from $|M|^2$ the terms that refer to the kinematically forbidden transition, one gets

$$\Sigma |M|^2 = -4Z^2 e^2 \left\{ \left[(p'_\alpha \tilde{P}^\alpha) - 2(p_\alpha \tilde{P}^\alpha) \right] - 8\pi^2 \frac{2GM}{b} \left[(p'_\alpha p^\alpha) - 2m^2 \right] \right\} \,. \tag{6.15}$$

The first set of square brackets in (6.15) represents the contribution of M_1 and coincides with (6.13). This supports the claim that the generalized momentum P_μ introduced in (6.4) leads to the correct value of the transition probability by the substitution of p_μ with P_μ in the field-free expression. The second set of square brackets represents the contribution of M_2.

The integration over d^3p' in (6.12) is performed by means of the identity

$$\int \frac{d^3p'}{2p_0'} = \int d^4p'\delta(p'^2 - m^2)\,,$$

while that over θ can be carried out by writing the on-shell condition for p' in the form

$$\delta(2|\mathbf{P}||\ell|\cos\theta - P^\alpha P_\alpha + 2P_0\ell_0 + m^2)\,. \tag{6.16}$$

One finds

$$W = \frac{2Z^2e^2}{p^2}\frac{GM}{b}\left\{m^2\ell_0^2 - \left[\frac{m^2\ell_0^2}{2} + \frac{\ell_0^3(p_0 - p)}{3}\right]\right\}. \tag{6.17}$$

The first term in (6.17) represents the contribution of M_1. The radiation spectrum is given by

$$\frac{dW}{d\ell_0} = \frac{2Z^2e^2\ell_0}{p^2}\left(\frac{GM}{b}\right)\left[m^2 - \ell_0\,(p_0 - p)\right]. \tag{6.18}$$

For $p > m$, (6.17) gives

$$W_{p>m} = Z^2e^2\frac{GM}{b}\frac{m^2\ell_0^2}{p^2}\,, \tag{6.19}$$

while it yields

$$W_{p<m} = Z^2e^2\frac{GM}{b}\frac{m\ell_0^2}{p^2}\left(\frac{m}{2} - \frac{\ell_0}{3}\right), \tag{6.20}$$

for $p < m$. Equation (6.16) and the condition $-1 \leq \cos\theta \leq 1$ require that for $p > m$

$$4\pi p\left(\frac{GM}{b}\right)\left(1 + \frac{m^2}{4p^2}\right) \leq \ell_0 \leq 4\pi p\left(\frac{GM}{b}\right)\left(\frac{4p^2}{m^2} + 3\right). \tag{6.21}$$

It follows from (6.21) that the hardest photons are emitted in the forward direction with energy

$$\ell_0 \sim 16\pi\left(\frac{GM}{b}\right)\frac{p^3}{m^2}\,,$$

and power

$$W_{p>m} \sim (16\pi Ze)^2\left(\frac{GM}{b}\right)^3\frac{p^4}{m^2}\,,$$

which takes its highest values in the neighbourhood of a black hole and for high values of p/m.

For $p < m$, one obtains

$$4\pi\left(\frac{GM}{b}\right)m\left(1 - \frac{p}{m}\right) \leq \ell_0 \leq 4\pi\left(\frac{GM}{b}\right)m\left(1 + \frac{p}{m}\right), \tag{6.22}$$

which reduces to the single value $\ell_0 \sim 4\pi m(GM/b)$ when $p \ll m$. In this case, the radiation is still in the forward direction and the power radiated

$$W_{p<m} \sim 8(\pi Ze)^2 \left(\frac{GM}{b}\right)^3 \frac{m^4}{p^2} \tag{6.23}$$

diverges for small values of p (infrared divergence). This divergence arises as a consequence of the finite energy resolution $\Delta\epsilon$ of the outgoing fermion. The process, as calculated, is in fact indistinguishable from that in which gravitons with energy $\leq \Delta\epsilon$ are also emitted and from processes in which vertex corrections are present (virtual gravitons emitted and reabsorbed by the external lines of Fig. 6.1). When these additional diagrams are calculated, all infrared divergences disappear [12]. In the particular case at hand, p in (6.23) is simply replaced by $4\pi GMp_0/b$, as requested by the external field approximation.

The results (6.17)–(6.20) ignore the back-reaction on the background space–time. It is shown below that this is not always negligible and an example is given of how a very small disturbance in the wave equation can grow rapidly and alter the background gravitational field.

The approximation procedure still holds true when $\Psi_0(x)$ satisfies more general equations [2,4,13,14]. With the addition of a dissipation term, the equation for Ψ_0 becomes

$$\left(\eta^{\mu\nu}\partial_\mu\partial_\nu + m^2 - 2m\sigma\,\partial_0\right)\Psi_0 = 0\,, \tag{6.24}$$

where [5]

$$\sigma = \alpha|\langle\Psi_0|\hat{T}|\Psi_0\rangle|^2 = \alpha\left(\frac{m}{p_0}\frac{GM}{2b}\right)^2,$$

and α is a dimensionless, arbitrary parameter, $0 \leq \alpha \leq 1$, that reflects the coupling strength of the dissipation term. When

$$\Psi_0(x) = e^{m}\sigma x_0\,\phi_0(x)$$

is substituted into (6.24), one gets

$$\left[\partial_0^2 - \partial_z^2 + m^2(1-\sigma^2)\right]\phi_0(x) = 0\,. \tag{6.25}$$

An example of a problem with similar behaviour is offered by a fluid heated from below. For small temperature gradients, the fluid conducts the heat, but as the gradient increases conduction is not sufficient to lead the heat away and the fluid starts to convect. In realistic problems, the exponential growth of Ψ_0 does not continue indefinitely, but is restricted at times $x_0 > \tau \equiv 1/m\sigma$ by nonlinearities or dispersive effects that may have been neglected initially.

The effect of the new solution Ψ_0 on W can be found as follows. First neglect the change

$$m \to m\sqrt{1-\sigma^2}$$

in W because in general $\sigma < 1$. The effect of the exponentially increasing term on (6.12) then amounts to the transformations

$$\delta(P_0 - p'_0 - \ell_0) \to \delta(-2i\sigma m + P_0 - p'_0 - \ell_0)$$

and

$$\frac{1}{p^2} \to \frac{1}{\beta^2(p_0 - 2i\sigma m)^2} \simeq \frac{p_0^2}{\beta^2[(p_0^2 - 4\sigma^2 m^2)^2 + 4p_0^2\sigma^2 m^2]},$$

where the relation $\beta = p/p_0$ has been used. W has therefore a resonance at

$$p_0 = 2\sigma m$$

of width $4\sigma^2 m^2$. Over times

$$x_0 > \tau = (\alpha m)^{-1}(m/p_0)^{-2}(GM/2b)^{-2}\text{GeV}^{-1},$$

the function $\Psi = \hat{T}\,\Psi_0$ increases exponentially until the compensating mechanisms mentioned above kick in. For a proton of energy $p_0 \sim 10$ GeV in the field of a canonical neutron star $\tau \sim 3.5 \cdot 10^{-21}\alpha^{-1}$ s. Considerably higher values of τ can, of course, be obtained for the lighter fermions. As Ψ grows, so does the energy momentum tensor associated with it and the gravitational field it generates, altering, in the process, the gravitational background.

Summarizing, external gravitational fields in radiative processes can be included in the calculation of a transition probability by simply replacing the momentum p_μ of a particle with its generalized version P_μ in the corresponding expression for the zero-field process. The example given refers to fermions, but the procedure can be extended to particles of different spins. An essential point is that the dispersion relations are altered by the external gravitational field and can be calculated if the corresponding wave equations can be solved to $O(\gamma_{\mu\nu})$, or higher [2,5–7]. The treatment of particle lines in Feynman diagrams therefore necessitates care when external gravitational fields are present. It follows, in particular, that kinematically forbidden processes like that of Fig. 6.1 become physical and their transition probabilities can be determined. The calculation of the gravitational contributions is greatly simplified and can be extended to higher order in $\gamma_{\mu\nu}$. The applications are not confined to fields of a Newtonian type, but extend to any gravitational fields. In this respect, the procedure presented goes beyond the results that apply to external electromagnetic potentials [8], not only because the gravitational field has, in general, ten components rather than just four, but also because time independence is not required in (6.7).

The procedure also yields transition amplitudes and decay rates that, to leading order, are linear in $\gamma_{\mu\nu}$ and can therefore be considerably larger than those normally studied in the literature. These results suggest that some particle decay processes in the neighbourhood of compact astrophysical objects, or in cosmology, may deserve a closer scrutiny.

Known theorems [8] then enable the calculation of transition amplitudes and power spectra for a photon that, after travelling through a gravitational field, produces a fermion–antifermion couple ($\gamma \rightarrow f + \bar{f}$) and for a fermion that, after crossing a gravitational field, interacts with an antifermion to produce a photon ($f + \bar{f} \rightarrow \gamma$).

It is also shown that the back-reaction of the fermion on the gravitational background need not be negligible. In fact, the addition of a small dissipation term in (6.24) can drastically transform the physical problem over a characteristic time τ in two respects: (i) the power radiated by the fermion acquires a resonant, narrow peak at $p_0 \sim 2\sigma m$ and (ii) Ψ grows exponentially and, in so doing, affects the background gravitational field via its associated energy–momentum tensor, until so allowed by the physical circumstances.

6.2 Spin Currents in Gravitational Fields

The realization that the flow of spin angular momentum can be separated from that of charge has stimulated intense interest in fundamental spin physics [15]. Spin control is an important issue in spin-based electronics [16] and in all phenomena essentially based on spin transport and dynamics [17].

The purpose of this section is to study the generation and control of spin currents by rotation and acceleration. In this context, a fundamental tool is the covariant Dirac equation [18,19] discussed in Chap. 1.

The first-order solutions of the covariant Dirac equation are of the form

$$\Psi(x) = \hat{T}(x)\psi_0(x)\,, \tag{6.26}$$

where, as usual, $\psi_0(x)$ is a solution of the free Dirac equation and $\hat{T}$ and Φ_T are given in Sect. 1.3.4 [5]. It is convenient to choose $\psi_0(x)$ in the form of plane waves.

The path integrals of Φ_G and Φ_T are taken along the classical world line of the fermion, specifically, but not necessarily an electron, starting from a reference point P. Only the path to $O(\gamma_{\mu\nu})$ needs to be known in the integrations indicated because (6.26) already are first-order solutions. The positive energy solutions of the free Dirac equation are given by

$$\psi_0(x) = u(\mathbf{k})e^{-ik_\alpha x^\alpha} = N\begin{pmatrix} \phi \\ \frac{\boldsymbol{\sigma}\cdot\mathbf{k}}{E+m}\phi \end{pmatrix} e^{-ik_\alpha x^\alpha}\,, \tag{6.27}$$

where

$$u^+u = 1\,, \quad \bar{u} = u^+\gamma^0\,, \quad u_1^+u_2 = u_2^+u_1 = 0\,,$$

and $N = \sqrt{\frac{E+m}{2E}}$. In addition, ϕ can take the forms ϕ_1 and ϕ_2 where

$$\phi_1 = \begin{pmatrix} 1 \\ 0 \end{pmatrix}\,, \quad \phi_2 = \begin{pmatrix} 0 \\ 1 \end{pmatrix}\,.$$

When acceleration and rotation are present, $\gamma_{\mu\nu}$ is given in [20] from which the spinorial connection can be calculated. From Sects. (1.3.4) and (1.3.5), one finds

$$\Gamma_i = 0\,, \quad \Gamma_0 = -\frac{1}{2}a_i\sigma^{\hat{0}\hat{i}} - \frac{1}{2}\left(\mathbf{\Omega}\cdot\sigma\right) I\,.$$

For electrons, u_1 corresponds to the choice $\phi = \phi_1$ and u_2 to $\phi = \phi_2$. One finds

$$u_1 = N\begin{pmatrix} 1 \\ 0 \\ \frac{k^3}{E+m} \\ \frac{k^1+ik^2}{E+m} \end{pmatrix}, \quad u_2 = N\begin{pmatrix} 0 \\ 1 \\ \frac{k^1-ik^2}{E+m} \\ \frac{-k^3}{E+m} \end{pmatrix},$$

which are not eigenspinors of the matrix $\Sigma^3 = \sigma^3 I$ whose eigenvalues represent the spin components in the z-direction, but become eigenspinors of Σ^3 when $k^1 = k^2 = 0$, or in the rest frame of the electron $\mathbf{k} = 0$.

The transfer of angular momentum between the external non-inertial field and the electron spins can be demonstrated using the spin current tensor [21]

$$S^{\rho\mu\nu} = \frac{1}{4im}\left[\left(\nabla^\rho\bar{\Psi}\right)\sigma^{\mu\nu}(x)\Psi - \bar{\Psi}\sigma^{\mu\nu}(x)\left(\nabla^\rho\Psi\right)\right], \tag{6.28}$$

which satisfies the conservation law $S^{\rho\mu\nu}{}_{,\rho} = 0$ when all $\gamma_{\alpha\beta}(x)$ vanish and yields in addition the expected result $S^{\rho\mu\nu} = \frac{1}{2}\bar{u}_0\sigma^{\hat{\mu}\hat{\nu}}u_0$ in the rest frame of the particle. Writing

$$\sigma^{\mu\nu}(x) \approx \sigma^{\hat{\mu}\hat{\nu}} + h^\mu_{\hat{\tau}}\sigma^{\hat{\tau}\hat{\nu}} + h^\nu_{\hat{\tau}}\sigma^{\hat{\mu}\hat{\tau}}\,,$$

using (6.26) in (6.28) and keeping only terms $O(\gamma_{\alpha\beta})$, one finds

$$S^{\rho\mu\nu} = \frac{1}{16im^3}\bar{u}_0\Big\{8im^2k^\rho\sigma^{\hat{\mu}\hat{\nu}} + 8imk^\rho h^{[\mu}_{\hat{\tau}}\sigma^{\hat{\tau}\hat{\nu}]} + \tag{6.29}$$

$$4imk^\rho\left(\Phi_{G,\alpha} + k_\sigma h^\sigma_{\hat{\alpha}}\right)\left\{\sigma^{\hat{\mu}\hat{\nu}}, \gamma^{\hat{\alpha}}\right\} - 8imk^\rho\Phi_G k^{[\mu}\gamma^{\hat{\nu}]} +$$

$$4mk^\rho k_\alpha\left[\sigma^{\hat{\mu}\hat{\nu}}, \left(\gamma^{\hat{\alpha}}\Phi_S - \gamma^{\hat{0}}\Phi_S^+\gamma^{\hat{0}}\gamma^{\hat{\alpha}}\right)\right] + 4m^2k^\rho\left[\sigma^{\hat{\mu}\hat{\nu}}, \left(\Phi_S - \gamma^{\hat{0}}\Phi_S^+\gamma^{\hat{0}}\right)\right] -$$

$$8m^2k^\rho h^0_{\hat{\alpha}}\left[\gamma^{\hat{0}}, \left[\sigma^{\hat{0}\hat{\alpha}}, \sigma^{\hat{\mu}\hat{\nu}}\right]\right] - 8im^2k_\sigma\left(\Gamma^\sigma_{\alpha\beta}\eta^{\beta\rho} + \partial^\rho h^\sigma_{\hat{\alpha}}\right)\eta^{\alpha[\mu}\gamma^{\hat{\nu}]} +$$

$$8im^2\partial^\rho\Phi_G\left(4m\sigma^{\hat{\mu}\hat{\nu}} - 2ik^{[\mu}\gamma^{\hat{\nu}]}\right) + 4im^2\gamma^{\hat{0}}\Gamma^{\rho+}\gamma^{\hat{0}}\left\{\left(\gamma^{\hat{\alpha}}k_\alpha + m\right), \sigma^{\hat{\mu}\hat{\nu}}\right\}\Gamma^\rho\Big\}u_0$$

where use has been made of the relation $\Phi_{G,\mu\nu} = k_\alpha\Gamma^\alpha_{\mu\nu}$. It is therefore possible to separate $S^{\rho\mu\nu}$ in inertial and non-inertial parts. The first term on the r.h.s. of (6.29) gives the usual result in the particle rest frame, when the external field vanishes. From (6.29), one finds

$$\partial_\rho S^{\rho\mu\nu} = \frac{1}{16im^3}\bar{u}_0 \left\{8imk^\rho \partial_\rho h^{[\mu}_{\hat{\tau}} \sigma^{\hat{\tau}\hat{\nu}]} - 8imk^\rho \Phi_{G,\rho} k^{[\mu}\gamma^{\hat{\nu}]} + \right. \tag{6.30}$$

$$4imk^\rho \left(k_\sigma \Gamma^\sigma_{\alpha\rho} + \partial_\rho h^\sigma_{\hat{\alpha}} k_\sigma\right)\left\{\sigma^{\hat{\mu}\hat{\nu}}, \gamma^{\hat{\alpha}}\right\} + 4mk^\rho k_\alpha \left[\sigma^{\hat{\mu}\hat{\nu}}, \left(\gamma^{\hat{\alpha}}\Gamma_\rho - \gamma^{\hat{0}}\Gamma^+_\rho \gamma^{\hat{0}}\gamma^{\hat{\alpha}}\right)\right] +$$

$$4m^2 k^\rho \left[\sigma^{\hat{\mu}\hat{\nu}}, \left(\Gamma_\rho - \gamma^{\hat{0}}\Gamma^+_\rho \gamma^{\hat{0}}\right)\right] + 8m^2 k^\rho \partial_\rho h^0_{\hat{\alpha}} \left[\gamma^{\hat{0}}, \left[\sigma^{\hat{0}\hat{\alpha}}, \sigma^{\hat{\mu}\hat{\nu}}\right]\right] -$$

$$8im^2 k_\sigma \partial^\rho \Gamma^\sigma_{\alpha\rho} \eta^{\alpha[\mu}\gamma^{\hat{\nu}]} + 8im^2 k_\sigma \Gamma^\sigma_{\rho\tau}\eta^{\tau\rho}\left(4m\sigma^{\hat{\mu}\hat{\nu}} - 2ik^{[\mu}\gamma^{\hat{\nu}]}\right) +$$

$$\left. 8im^2 k^\alpha \Gamma^\rho_{\alpha\rho}\sigma^{\hat{\mu}\hat{\nu}} + 8im^2 k^\rho \Gamma^\mu_{\alpha\rho}\sigma^{\hat{\alpha}\hat{\nu}} - 8im^2 k^\rho \Gamma^\nu_{\alpha\rho}\sigma^{\hat{\alpha}\hat{\mu}}\right\} u_0 ,$$

where terms containing $\Gamma_{0,0} = 0$ and $\partial_\alpha \partial_\beta h^\mu_{\hat{\nu}} = 0$ have been eliminated. It follows from (6.30) that the external field invalidates the conservation law and that, therefore, there is continual interchange between spin and orbital angular momentum. The result is entirely similar to that observed for external electromagnetic fields [21]. This essentially proves that non-inertial fields can be used in principle to generate and control spin currents. Because of the complexity of the equations, detailed studies of spin currents require lengthy calculations. In the rest frame of the particle, when $\mathbf{\Omega} = (0, 0, \Omega)$, one finds

$$\partial_\rho S^{\rho\mu\nu} = \partial_i S^{i12} = \frac{1}{2}\left(\Gamma^\rho_{0\rho} + \Gamma^1_{10} + \Gamma^2_{20}\right)\bar{u}_0 \sigma^{\hat{1}\hat{2}} u_0 = \frac{E+m}{2E}\frac{\Omega a_2 x}{1 + \mathbf{a}\cdot\mathbf{x}}, \tag{6.31}$$

and $\partial_i S^{i13} = \partial_i S^{i23} = 0$. In (6.31) u_0 corresponds to u_1. The violation of the conservation law is thus due to the direct coupling of the non-inertial field to the particle's spin current. Conservation is restored if either Ω, or a_2, or both vanish.

It is also useful to consider the actual spin motion. Transfer of angular momentum between external and non-inertial fields takes place when the operator $\hat{T}$ has some non-diagonal matrix elements. If in fact at time $t = 0$ a beam of electrons is entirely of the u_2 variety, at time t the fraction of u_1 is $|\langle u_1|\hat{T}|u_2\rangle|^2$. It is more convenient to write the last expression in the form

$$P_{2\to 1} = \left|\langle u_1|\hat{T}|u_2\rangle\right|^2 = \left|\int_{\lambda_0}^{\lambda}\langle u_1|\dot{x}^\mu \hat{T}_{,\mu}|u_2\rangle d\lambda\right|^2 , \tag{6.32}$$

where $\dot{x}^\mu = k^\mu/m$ and λ is an affine parameter along the electron world line. From [5]

$$\hat{T}_{,\nu} = \frac{1}{2m}\left\{h^\mu_{\hat{\alpha},\nu}\gamma^{\hat{\alpha}}k_\mu + \gamma^{\hat{\mu}}\Phi_{G,\mu\nu} - 2im\left(\Phi_{G,\nu} + \Gamma_\nu - eA_\nu\right)\right\} , \tag{6.33}$$

one can see that

$$\langle u_1|\frac{k^\nu}{m}\hat{T}_{,\nu}|u_2\rangle = -i\frac{k^0}{m}\langle u_1|\Gamma_0|u_2\rangle = -i\frac{k^0}{m}\langle u_1|\left\{-\frac{1}{2}a_i\sigma^{\hat{0}\hat{i}} - \frac{1}{2}\Omega_i\sigma^i I\right\}|u_2\rangle, \tag{6.34}$$

while the Mashhoon coupling $H_M = -\mathbf{\Omega}\cdot\mathbf{s}$ [22], where $\mathbf{s} = \frac{\sigma}{2}$, and the interaction term $\frac{1}{2}a_i\sigma^{\hat{0}\hat{i}} = \frac{i}{2}a_i\alpha^{\hat{i}}$ gives the first-order equation of motion [23]

$$\frac{ds}{dt} = \mathbf{s}\times(\mathbf{\Omega} + \mathbf{v}\times\mathbf{a})\ , \tag{6.35}$$

which is useful in visualizing the spin motion under the action of rotation and acceleration. Notice that A_μ does not contribute to (6.34) because $\langle u_1|u_2\rangle = 0$ and that the terms $ie(h^{\mu}_{\hat{\alpha}}\gamma^{\hat{\alpha}}k_\mu + \gamma^{\hat{\mu}}\Phi_{G,\mu})A_\nu\frac{k^\nu}{m}$ drop out on account of $\langle u_1|\gamma^{\hat{\mu}}|u_2\rangle = 0$. No mixed effects of first order in rotation or acceleration and first order in the electromagnetic field are therefore present in this calculation. This applies to all terms containing the magnetic field **B**, like the Zeeman term, and electric fields, like the spin–orbit interaction, that are present in the lowest order Dirac Hamiltonian that can be derived from (6.55). To $O(\gamma_{\mu\nu})$, contributions to (6.34) from the electromagnetic field are present in the actual determination of the electron's path, as stated above.

From (6.34), one obtains

$$\begin{aligned}\frac{2m}{iE}\langle u_1|\frac{k^\nu}{m}\hat{T}_{,\nu}|u_2\rangle &= -i\frac{k^3}{E}a_1 - \frac{k^3}{E}a_2 + i\frac{k^1 - ik^2}{E}a_3 \\ &+ \Omega^3\frac{k^3}{E}\frac{-k^1 + ik^2}{E+m} \\ &+ \Omega^1\frac{E+m}{2E}\left(1 + \frac{(k^3)^2}{(E+m)^2} - \frac{(k^1 - ik^2)^2}{(E+m)^2}\right) \\ &- i\Omega^2\frac{E+m}{2E}\left(1 + \frac{(k^3)^2}{(E+m)^2} + \frac{(k^1 - ik^2)^2}{(E+m)^2}\right) \equiv A_{12}\,,\end{aligned} \tag{6.36}$$

where $k^0 \equiv E$. The parameter k_μ corresponds to the electron four momentum when $\mathbf{\Omega} = 0$ and $\mathbf{a} = 0$. No conversion spin-up to spin-down is possible when $A_{12} = 0$.

Some general conclusions can be drawn from (6.36). (i) If $k^3 = 0$, the particles move in the (x, y)-plane. If, in addition, $\Omega^1 = \Omega^2 = 0$, then $A_{12} \neq 0$ only if $a_3 \neq 0$. (ii) If, however, **a** is also due to rotation, the conditions $\Omega_1 = \Omega_2 = 0$ imply $a_3 = 0$ and therefore $A_{12} = 0$. This is the relevant case of motion in the (x, y)-plane with rotation along an axis perpendicular to it. One cannot therefore have a rotation-induced spin current in this instance. (iii) Even for $\mathbf{k} = 0$ one can have $A_{12} \neq 0$ if one of Ω_1 and Ω_2 does not vanish. This is a direct consequence of Mashhoon's spin-rotation interaction contained in the covariant Dirac equation [20,22].

Rotation and acceleration generate and control spin currents. This follows from the covariant Dirac equation and its exact solutions to $O(\gamma_{\mu\nu})$. The transition amplitude for the conversion spin-up to spin-down is proportional to A_{12} and is expressed as a function of $\mathbf{\Omega}$, $\mathbf{a}$ and of the electron four momentum k_μ before the onset of rotation and acceleration. The same expression suggests criteria for the generation of spin currents and the transfer of momentum and angular momentum to them. No energy can, of course, be transferred from the non-inertial fields to the spin currents when $\gamma_{\mu\nu}$ is time independent.

An analysis of the effect of mechanical rotation on spin currents has also been given in [24–26]. Additional non-inertial effects in quantum systems with cosmological applications have been considered by Bakke [27].

6.3 Vortices

Symmetry breaking in quantum many-body systems gives rise to macroscopic objects like vortices in superconductors, dislocations in crystals, and domain walls in ferromagnets. These structures normally appear in a quantum context, but behave classically. They are properties of matter, in the form other than particles, that emerge from a quantum background when quantum fluctuations become negligible.

Similar phenomena may occur in quantum gravity, still far from the essential quantum regime that is supposed to take over at Planck's length.

The first-order solution of the Klein–Gordon equation in its minimal coupling form and after applying the Lanczos–DeDonder condition (1.39) is given by (1.37) and (1.38). The expression Φ_G that appears in (1.37) can be rewritten in the form

$$\hat{\Phi}_G(x) = \int_P^x dz^\lambda \hat{K}_\lambda(z, x)\,. \tag{6.37}$$

The transformation (1.37) that makes the ground state of the system space–time dependent results in a breakdown of symmetry.

Choosing a plane wave for ϕ_0 so that the wave vector k_α satisfies the condition $k_\alpha k^\alpha = m^2$, then

$$K_\lambda(z, x) = -\frac{1}{2}\left[\left(\gamma_{\alpha\lambda,\beta}(z) - \gamma_{\beta\lambda,\alpha}(z)\right)\left(x^\alpha - z^\alpha\right) - \gamma_{\beta\lambda}(z)\right]k^\beta\,. \tag{6.38}$$

Notice that K_λ contains information about the particles with which gravity interacts through the momentum k_α of ϕ_0. The quanta of K_λ can be called quasiparticles this being the notion that explains the properties of fields and particles that are affected by the interaction with other particles and media. As the free particles feel the gravitational field, the system ground state changes and evolves towards a lower equilibrium configuration. The loss of symmetry is determined by the curvature of space–time. When ϕ_0 is transported along a closed path Γ bounding a small surface $df^{\alpha\beta}$ in curved space–time, it changes by

$$\delta\phi \sim \frac{i}{4}\int_\Sigma df^{\beta\delta} R_{\mu\nu\beta\delta} J^{\mu\nu}\phi_0\,.$$

The rotationally invariant ϕ_0 therefore acquires a privileged direction. The direction of rotation associated with $J_{\mu\nu}$ breaks the original symmetry of the vacuum. The process is known as boson condensation [28]. The effects of boson condensation in EFA have been studied in [29,30] and amount to a temperature-dependent correlation length and to momentum oscillations about an average momentum that are similar to spin waves and have a Planck distribution. The condensation phenomena also lead to gravitational analogues of the laws of Curie and Bloch and to a critical temperature below which vortices form in pairs.

By differentiating (6.38) with respect to z^α, one finds [31]

$$\tilde{F}_{\mu\lambda}(z,x) \equiv K_{\lambda,\mu}(z,x) - K_{\mu,\lambda}(z,x) = R_{\mu\lambda\alpha\beta}(z) J^{\alpha\beta}\,, \tag{6.39}$$

where $R_{\alpha\beta\lambda\mu}(z)$ is the linearized Riemann tensor (1.25) that satisfies the identity

$$R_{\mu\nu\sigma\tau} + R_{\nu\sigma\mu\tau} + R_{\sigma\mu\nu\tau} = 0\,,$$

and

$$J^{\alpha\beta} = \frac{1}{2}\left[\left(x^\alpha - z^\alpha\right)k^\beta - k^\alpha\left(x^\beta - z^\beta\right)\right]$$

is the angular momentum about the base point x^α. The Maxwell-type equations

$$\tilde{F}_{\mu\lambda,\sigma} + \tilde{F}_{\lambda\sigma,\mu} + \tilde{F}_{\sigma\mu,\lambda} = 0 \tag{6.40}$$

and

$$\tilde{F}^{\mu\lambda}{}_{,\lambda} \equiv -j^\mu = \left(R^{\mu\lambda}{}_{\alpha\beta}J^{\alpha\beta}\right)_{,\lambda} = R^{\mu\lambda}{}_{\alpha\beta,\lambda}\left(x^\alpha - z^\alpha\right)k^\beta + R^\mu{}_\beta k^\beta \tag{6.41}$$

can be obtained from (6.39) using the Bianchi identities

$$R_{\mu\nu\sigma\tau,\rho} + R_{\mu\nu\tau\rho,\sigma} + R_{\mu\nu\rho\sigma,\tau} = 0\,.$$

The current j^μ satisfies the conservation law $j^\mu{}_{,\mu} = 0$. Equations (6.40) and (6.41) are identities and do not represent additional constraints on $\gamma_{\mu\nu}$.

The vector K_λ is non-vanishing only on surfaces $\tilde{F}_{\mu\nu}$ that satisfy (6.40) and (6.41) and represent the vortical structures generated by Φ_G. At a point z_α along the path, one finds

$$\frac{\partial\Phi_g(z)}{\partial z^\sigma} \equiv -\frac{1}{2}\left[\left(\gamma_{\alpha\sigma,\beta}(z) - \gamma_{\beta\sigma,\alpha}(z)\right)\left(x^\alpha - z^\alpha\right) - \gamma_{\beta\sigma}(z)\right]k^\beta = K_\sigma(z)\,, \tag{6.42}$$

and

$$\frac{\partial^2\Phi_g(z)}{\partial z^\tau \partial z^\sigma} - \frac{\partial^2\Phi_g(z)}{\partial z^\sigma \partial z^\tau} = R_{\alpha\beta\sigma\tau}\left(x^\alpha - z^\alpha\right)k^\beta \equiv \left[\partial z_\tau, \partial z_\sigma\right]\Phi_g(z) = \tilde{F}_{\tau\sigma}(z)\,. \tag{6.43}$$

It follows from (6.43) that Φ_g is not single valued and that, after a gauge transformation, K_α satisfies the equations

$$\partial_\alpha K^\alpha = \frac{\partial^2 \Phi_g}{\partial z_\sigma \partial z^\sigma} = 0 \tag{6.44}$$

and

$$\partial^2 K_\lambda = -\frac{k^\beta}{2}\left[\left(\partial^2(\gamma_{\alpha\lambda,\beta}) - \partial^2(\gamma_{\beta\lambda,\alpha})\right)\left(x^\alpha - z^\alpha\right) - \partial^2 \gamma_{\beta\lambda}\right] \tag{6.45}$$

identically, while the equation

$$\left[\partial z_\mu, \partial z_\nu\right] \partial z_\alpha \Phi_g = -\left(\tilde{F}_{\mu\nu,\alpha} + \tilde{F}_{\alpha\mu,\nu} + \tilde{F}_{\mu\alpha,\nu}\right) = 0 \tag{6.46}$$

holds everywhere. The potential K_α is therefore regular everywhere, which is physically desirable, but Φ_g has a Berry wave singularity [32]. There must then be closed paths embracing the singularities along which the particle wave function must be made single valued by means of appropriate quantization conditions [30]. The change experienced by K_α amounts to

$$\oint_\Gamma dz_\sigma K^\sigma(z) = \int_\Sigma df^{\sigma\tau} \tilde{F}_{\sigma\tau} = \int_\Sigma ds^{\sigma\tau} R_{\sigma\tau\alpha\beta} J^{\alpha\beta} = 2\pi n\hbar,$$

which confers some vortical structure to space–time.

It also follows from (6.43) that $\tilde{F}_{\mu\nu}$ is a vortex along which the scalar particles are dragged with acceleration

$$\frac{d^2 z_\mu}{ds^2} = u^\nu \left(u_{\mu,\nu} - u_{\nu,\mu} - R_{\mu\nu\alpha\beta}\left(x^\alpha - z^\alpha\right) u^\beta\right), \tag{6.47}$$

and relative acceleration

$$\frac{d^2(x_\mu - z_\mu)}{ds^2} = \tilde{F}_{\mu\lambda} u^\lambda = R_{\mu\beta\lambda\alpha}\left(x^\alpha - z^\alpha\right) u^\beta u^\lambda, \tag{6.48}$$

in agreement with the equation of geodesic deviation [30]. Notice that in (6.47) the vorticity is entirely due to $R_{\mu\nu\alpha\beta} J^{\alpha\beta}$ and that $\frac{d^2 z_\mu}{ds^2} = 0$ when the motion is irrotational. This also applies when $R_{\mu\nu\alpha\beta} = 0$, in which case the vortices do not develop. Similarly, vortices do not form if $k^\alpha = 0$. Each gravitational field produces a distinct vortex whose equations are (6.40) and (6.41), the vortex dynamics is given by (6.47) and (6.48) and the topology of the object is supplied by Φ_g. Though their origin resides in a quantum wave equation, the vortices generated are purely classical because $\gamma_{\mu\nu}$, K_λ, and $\tilde{F}_{\alpha\beta}$ are classical and the particles interact with gravity as classical particles do. In addition, ϕ and ϕ_0 coexist with the vortices generated by Φ_G in the ground state. The field $\tilde{F}_{\mu\nu}$ emerges as a property of gravitation when this interacts with particles described by wave equations in the external field approximation. Its range is that of $\gamma_{\mu\nu}$. $\tilde{F}_{\alpha\beta}$ vanishes on the line $x^\alpha - z^\alpha = 0$ along which K_λ can also be eliminated by a gauge transformation. One can say that in this case the line is entirely occupied by ϕ_0. Obviously, $\Phi_G = 0$ on the nodal lines of ϕ where it looses its meaning. Notice that the left-hand side of (6.39) can also be replaced by its dual. This is equivalent to interchanging the "magnetic" with the "electric" components of $R_{\mu\nu\alpha\beta}$ and the corresponding vortex types.

The simplest possible Lagrangian in which the features just discussed can be accommodated is [28]

$$\mathcal{L} = -\frac{1}{4}\tilde{F}_{\alpha\beta}\tilde{F}^{\alpha\beta} + \left[\left(\partial_\mu - iK_\mu\right)\phi\right]^* \left[\left(\partial^\mu + iK^\mu\right)\phi\right] - \mu^2\phi^*\phi\,, \tag{6.49}$$

where $\mu^2 < 0$. The second term of $\mathcal{L}$ contains the first-order gravitational interaction $\gamma_{\mu\nu}[(\partial^\mu - iK^\mu)\phi]^*[(\partial^\nu + iK^\nu)\phi] \sim -\gamma_{\mu\nu}\partial^\mu\partial^\nu\phi_0$ met above. By varying $\mathcal{L}$ with respect to ϕ^* and by applying a gauge transformation to K_α, one finds to $O(\gamma_{\mu\nu})$,

$$\left[\partial^2 + m^2 + \gamma_{\mu\nu}\partial^\mu\partial^\nu\right]\phi(x) \simeq 0\,, \tag{6.50}$$

and $-\mu^2$ has now been changed into $m^2 > 0$ because the Goldstone boson has disappeared, the remaining boson is real [28] and so must be its mass. Equation (6.50) is identical to (1.36) and its solution is still represented by Eq. (1.37). However, a variation of $\mathcal{L}$ with respect to K_α now gives

$$\partial_\nu\tilde{F}^{\mu\nu} = \tilde{J}^\mu = i\left[\left(\phi^*\partial^\mu\phi\right) - \left(\partial^\mu\phi^*\right)\phi\right] - 2K^\mu\phi^*\phi \tag{6.51}$$

which, on using (1.37) and a gauge transformation, leads to the field equation

$$\partial^2 K_\mu + 2K_\mu\phi^2 = 0 \tag{6.52}$$

that shows that K_μ has acquired a mass. By using the expansion $\phi = v + \rho(x)/\sqrt{2}$, the mass of K_μ is v and its range $\sim v^{-1}$. Any metrical theory of gravity selected remains valid at distances greater than v^{-1}, but not so near, or below v^{-1}. The shielding current in Eq. (6.52) determines a situation analogous to that of vortices of normal electrons inside type-II superconductors where the electron normal phase is surrounded by the condensed, superconducting phase. The fundamental difference from the approach followed before the introduction of $\mathcal{L}$ is represented by (6.52) that now becomes a constraint on K_λ. It can be satisfied by requiring that $(\partial^2 + v^2)\gamma_{\alpha\beta} = 0$. No other changes are necessary. On the other hand, this condition can be applied independently of $\mathcal{L}$. $\tilde{F}_{\mu\nu}$ again vanishes when $z^\alpha - x^\alpha = 0$, which indicates that the line $z^\alpha - x^\alpha = 0$ can only be occupied by the normal phase. As before, the field $\tilde{F}_{\mu\nu}$ is classical and emerges as a property of gravitation when it interacts with quantum matter. The range of interaction can obviously be very short if v is large. Shielding now produces vortices in which the normal phase is trapped. The screening length $\sim v^{-1}$ must be small in order to prevent macroscopic violations of gravity's universal law of attraction and the instability of any particle, or system of particles, whose internal mechanical behaviour involves inertial forces. The only known scalar particles fitting the requirements belong to the Higgs boson family, or to new, undiscovered particles if indeed v^{-1} must be small.

The models of this section can conceal matter in vortical structures that interact only gravitationally with the rest of the universe over long-, or short-range distances, or both. They may be of interest in the study of dark matter and dark energy.

6.3.1 Spin-$\frac{1}{2}$ Fermions

As shown in Chap. 1, the first-order solution of the covariant Dirac equation can be written in the form

$$\Psi(x) = -\frac{1}{2m}\left(-i\gamma^{\mu}(x)\mathcal{D}_{\mu} - m\right)e^{-i\Phi_T}\Psi_0(x)\,. \tag{6.53}$$

It is shown in [2,5] that the solution requires that (1.36) be also solved. This accounts for the presence of Φ_T in (6.53).

Consider the question whether vortices can be created by rotation either in laboratory conditions, or in the vicinity of an astrophysical source. In the first instance, rotation can be introduced by means of the Thirring metric [33] that describes the field of a shell of mass M and radius R rotating with angular velocity ω about the z-axis. The components of interest are J_{12}, R_{1212}, and $\gamma_{12} = 16\pi GM\omega^2 xy/5R$. Taking the limit $2GM/R \approx 1$ is considered appropriate when the spherical shell refers to the whole universe. This is in fact equivalent to assuming that the universe is rotating relative to the particles, which is consistent with Mach views. The limit can be also derived from an exact solution of Einstein equations [34]. One finds

$$\Phi_T = (1/2)\oint_{\sigma} R_{1212}J^{12}d\sigma^{12} = (\omega^2/5)\oint_{\sigma} J^{12}d\sigma^{12}\,,$$

where σ is the surface bounded by the particle path and J^{12} also contains the spin contribution because the spin connection satisfies Equation (1.74) [2,35] and

$$[\mathcal{D}_{\mu}, \mathcal{D}_{\nu}] = -\frac{i}{4}\sigma^{\alpha\beta}(x)R_{\alpha\beta\mu\nu}\,. \tag{6.54}$$

In the case of rotating astrophysical sources, one can use the Lense–Thirring metric [33]. The components of interest are again J_{12} and R_{1210}, R_{1220}, R_{1230} and the metric components to consider are $\gamma_{0i} = (4\alpha R^3\omega/5r^3)(y, -x, 0)$, where $\alpha = GM/R$. When, close to the source, $r \approx R$, one has

$$R_{1210} = -\frac{\alpha\omega x}{R^2}\,, \quad R_{1220} = \frac{\alpha\omega y}{R^2}\,, \quad R_{1230} = \alpha\omega\left(-\frac{3z}{R^2} + \frac{5z^3}{R^4}\right)\,,$$

from which it follows that

$$\Phi_T = \oint_\sigma R_{12\alpha\beta} J^{12} d\sigma_{\alpha\beta}$$

and, as above, J^{12} also contains the *spin* contribution to the total angular momentum. By applying the external field approximation to the covariant Dirac equation and by using the Thirring and Lense–Thirring metrics, relativistic fermion vortices can, in principle, be produced by rotation in laboratory and astrophysical conditions.

The problem discussed in this section bears some resemblance to the question, also discussed in the literature, whether vortices can be produced in the laboratory with non-relativistic and relativistic electrons. The vortices discussed in this section are fully relativistic and have their origin in the phase singularities introduced by Berry [32] and Nye and Berry [36].

There are practical applications in which electron vortices can be useful [37,38], in particular, in the study of nanoparticles and their magnetic properties [39]. Non-relativistic electrons are, in a sense, easier to deal with because their orbital and spin angular momenta are separately conserved. As seen in Sect. 6.2, this is not the case for fully relativistic electrons for which only the total angular momentum is conserved [40,41]. In Sect. 6.2, the vortical motion is, in fact, transferred to the fermions by gravitational fields through the equation of deviation. This occurs in astrophysical contexts, but seems difficult to achieve with wave packets in the laboratory without using electromagnetic fields that might interfere with other objectives of the research.

A quantum field theory of vortical structures can also be found in the literature [42,43].

6.4 Zitterbewegung and Gravitational Berry Phase

It is shown in this section that the gravitational Berry phase also gives rise to a field-dependent Zitterbewegung (ZB) in the propagation of particles in a gravitational background [44]. This is a consequence of the fact that Berry phases mix states of positive and negative energies in the propagation of fermions and bosons. The results are valid in any reference frame and to any order of approximation in the metric deviation.

Consider the covariant Dirac equation

$$[i\gamma^\mu(x)\mathcal{D}_\mu - m]\Psi(x) = 0\,, \tag{6.55}$$

and write the first-order solutions of (6.55) in the form (6.53).

In most applications, $\psi_0(x)$ is represented by a positive energy solution

$$\psi(x) = u(\mathbf{k})e^{-ik_\mu x^\mu}\,.$$

However, the influence of negative energy solutions

$$\psi^{(1)}(x) = v(\mathbf{k})e^{ik_\mu x^\mu}$$

cannot be neglected because the wave functions $\psi(x)$ by themselves do not form a complete set. A relationship between $\Psi(x)$ and $\Psi^{(1)}(x) = \hat{T}_1\psi^{(1)}(x)$ must therefore be found. The spin-up (↑) and spin-down (↓) components of the spinors u and v obey the well-known equations

$$u_\downarrow = \gamma^5 v_\uparrow\,, \qquad v_\downarrow = \gamma^5 u_\uparrow\,. \tag{6.56}$$

The required relation between $\Psi(x)$ and $\Psi^{(1)}(x)$ follows from the replacement of $\psi_0(x)$ with $\gamma^5\psi_0(x)$. If, in fact,

$$\Psi(x) = e^{-ik_\mu x^\mu}\hat{T}u$$

is a solution of (6.55), it then follows from the relations

$$\left\{\gamma^5, \gamma^{\hat{\mu}}\right\} = 0, \quad \sigma^{\hat{\alpha}\hat{\beta}} = \frac{i}{2}[\gamma^{\hat{\alpha}}, \gamma^{\hat{\beta}}], \quad [\gamma^5, \Gamma^\mu] = 0\,,$$

$$\gamma^\mu(x) = e^\mu_{\hat{\alpha}}(x)\gamma^{\hat{\alpha}}\,, \quad \Gamma_\mu(x) = -\frac{1}{4}\sigma^{\hat{\alpha}\hat{\beta}}e^\nu_{\hat{\alpha}}e_{\nu\hat{\beta};\,\mu}$$

that $\Psi^{(1)}(x) = e^{ik_\mu x^\mu}\hat{T}_1 v$ also is a solution of (6.55) and $\hat{T}_1 = \gamma^5\hat{T}\gamma^5$. It is useful to further isolate the gravitational contribution in the vierbein components by writing $e^\mu_{\hat{\alpha}} \simeq \delta^\mu_{\hat{\alpha}} + h^\mu_{\hat{\alpha}}$, which leads to

$$\begin{aligned}\hat{T} &= \frac{1}{2m}\left\{(1 - i\Phi_G)\left(m + \gamma^{\hat{\alpha}}k_\alpha\right) - i\left(m + \gamma^{\hat{\alpha}}k_\alpha\right)\Phi_S + \left(k_\beta h^\beta_{\hat{\alpha}} + \Phi_{G,\alpha}\right)\gamma^{\hat{\alpha}}\right\}\\ &\equiv \hat{T}_0 + \hat{T}_G\,,\end{aligned} \tag{6.57}$$

where

$$\hat{T}_0 \equiv \frac{1}{2m}\left(m + \gamma^{\hat{\alpha}}k_\alpha\right)$$

and $\hat{T}_G$ contains the gravitational corrections.

The gravitational field mixes the positive and negative energy solutions of the free Dirac equation. In fact, the eigenstates $U^\pm = 1/\sqrt{2}(u \pm v)$ of γ^5 and the eigenstates u and v of $\hat{T}_0$ are not the same and $\hat{T}$, $\hat{T}_1$ mix u and v. The mixing is carried out by $\hat{T}_G$ and $\hat{T}_{1G}$ which are entirely due to Berry phase.

The state of a fermion in a gravitational field can be written in the form

$$|\Phi(t)\rangle = \alpha(t)|\psi(t)\rangle + \beta(t)|\psi^{(1)}(t)\rangle = \alpha_0\hat{T}(t)|\psi(t)\rangle + \beta_0\hat{T}_1(t)|\psi^{(1)}(t)\rangle\,, \tag{6.58}$$

where $|\alpha_0|^2 + |\beta_0|^2 = 1$, from which one obtains

$$\alpha(t) = \langle\psi|\Phi(t)\rangle = \alpha_0\langle\psi|\hat{T}|\psi\rangle + \beta_0\langle\psi|\hat{T}_1|\psi^{(1)}\rangle\,, \tag{6.59}$$

$$\beta(t) = \langle\psi^{(1)}|\Phi(t)\rangle = \alpha_0\langle\psi^{(1)}|\hat{T}|\psi\rangle + \beta_0\langle\psi^{(1)}|\hat{T}_1|\psi^{(1)}\rangle\,.$$

If at $t = 0$ the gravitational field is not present, then $\hat{T}_G = 0$, $\hat{T}_{1G} = 0$ and $\alpha(0) \equiv \alpha_0$, $\beta(0) \equiv \beta_0$. It follows from (6.59) that, as the system propagates in a gravitational field, shifts from $|\psi(t)\rangle$ to $|\psi^{(1)}(t)\rangle$ produce oscillations. Thus, the geometrical structure of space–time, represented by gravity, affects Hilbert space by producing oscillations between the positive and negative energy states.

The presence of an electromagnetic field can be accommodated [5] by adding the term $q A_\alpha$, where q is the charge of the particle, to $\Phi_{G,\alpha}$ in $\hat{T}$ and $\hat{T}_1$. The relationship between external electromagnetic fields and ZB has been investigated extensively by Feschbach and Villars [45] for both Dirac and Klein–Gordon equations.

In order to obtain the transition probabilities $|\alpha(t)|^2$, $|\beta(t)|^2$ from (6.59) in a concrete case, one can choose for simplicity

$$\psi(x) = f_{0,R} e^{-ik_\alpha x^\alpha} = \sqrt{\frac{E+m}{2m}} \begin{pmatrix} f_R \\ \frac{\sigma^3 k}{E+m} f_R \end{pmatrix} e^{-ik_\alpha x^\alpha}, \tag{6.60}$$

where f_R is the positive helicity eigenvector. The normalizations are $\langle\psi|\psi\rangle = 1$, where $\langle\psi| = \langle\psi^\dagger|\gamma^{\hat{0}}$, $\langle\psi^{(1)}|\psi^{(1)}\rangle = -1$, and $\langle\psi|\psi^{(1)}\rangle = \langle\psi^{(1)}|\psi\rangle = 0$. In addition, one needs explicit expressions of the metric components for the purpose of calculating $\hat{T}$ and $\hat{T}_1$. The choice of the metric [10]

$$\gamma_{00} = 2\phi, \quad \gamma_{ij} = 2\phi\delta_{ij}, \tag{6.61}$$

where $\phi = -\frac{GM}{r}$, and M, R are mass and radius of the source, is again dictated by simplicity. The tetrad components to order $O(\gamma_{\mu\nu})$ are given by

$$e^0_{\hat{i}} = 0, \quad e^0_{\hat{0}} = 1 - \phi, \quad e^i_{\hat{0}} = 0, \quad e^l_{\hat{k}} = (1+\phi)\,\delta^l_k. \tag{6.62}$$

Without loss of generality, one may consider particles starting from $z = -\infty$, and propagating along $x = b \geq R$, $y = 0$ in the field of the gravitational source and set $k^3 \equiv k$ and $k^0 \equiv E$.

Returning to (6.59), if originally the system is in a positive energy state, then $\alpha_0 = 1$, $\beta_0 = 0$, $|\Phi(t)\rangle = \hat{T}|\psi\rangle$ and from (6.58) and $\Phi_{G,3} = (E^2/k + k)\phi$ one gets

$$\beta(t) = \frac{e^{-2iq_\alpha x^\alpha}}{2m} \left\{ -\langle\psi^{(1)}| \left[E h^0_{\hat{0}} \gamma^{\hat{0}} + \left(-k h^3_{\hat{3}} + \left(\frac{E^2}{k} + k \right) \phi(z) \right) \gamma^{\hat{3}} \right] |\psi\rangle \right\}, \tag{6.63}$$

where $q_0 \equiv E$ because the field does not depend on time, hence energy is conserved, and $q_i \equiv k_i^{(i)} - k_i^{(f)}$. The first two terms in (6.63) are due to Γ_μ and refer to Φ_S. The remaining two terms come from $\Phi_{G,3}$ and also are Berry phase contributions. Thus, according to (6.59) and (6.63), the propagation of the particle has two overlapping components: one in which the state of the particle does not change, the other in which oscillations take place from and to energy states of opposite sign with a frequency $2E/\hbar$ in ordinary units. This is at least as large as the ZB frequency $2m$. The particle therefore behaves as if it were trying to conserve energy–momentum and angular momentum during its propagation. The presence of the gravitational Berry phase translates into a ZB that vanishes when there is no gravity acting on the particle and is therefore due to a real force, as pointed out in [45] for the case of an external electromagnetic field. Because the approach is covariant, the result holds true in any frame of reference. Moreover, the non-local potential $K_\lambda(x, x_0)$ can be calculated to any order, meaning that a ZB also exists at any order.

The transition amplitude $\langle\psi|\hat{T}|\psi\rangle$ is calculated using the relation

$$\langle\psi|\hat{T}|\psi\rangle = \int_{\lambda_0}^{\lambda} \langle\psi|\dot{x}^\mu \partial_\mu \hat{T}|\psi\rangle d\lambda\,,$$

where $\dot{x}^\mu = k^\mu/m$ and λ is an affine parameter along the particle world line. The calculation is outlined in [5].

The probability of the transition $\psi \rightarrow \psi^{(1)}$ follows from (6.63) and is

$$P_{\psi\rightarrow\psi^{(1)}} = |\beta(t)|^2 = \left[\frac{1}{2m^2}(k^2 - \frac{E^3}{k})\right]^2 \phi^2(z)\,. \tag{6.64}$$

If $\alpha_0 = 0$, $\beta_0 = 1$, then $|\alpha(t)|^2$ represents the probability for the inverse process $\psi^{(1)} \rightarrow \psi$. One finds

$$\begin{aligned} P_{\psi^{(1)}\rightarrow\psi} = |\alpha(t)|^2 = |\langle\psi|\hat{T}|\psi^{(1)}\rangle|^2 = |\langle\psi|\gamma^5\hat{T}_1\gamma^5|\psi^{(1)}\rangle|^2 = \\ = |\langle\psi^{(1)}|\hat{T}_1|\psi\rangle|^2 = P_{\psi\rightarrow\psi^{(1)}}\,. \end{aligned} \tag{6.65}$$

According to (6.64) and (6.65), the transitions generated by the simple metric (6.61) proceed in both directions with the same probability $\sim 9\phi^2/16$ when $k \gg mc$.

As mentioned above, an external electromagnetic field can be introduced by simply adding the corresponding Berry phase to (6.63). The additional term in curly brackets is therefore $\langle\psi^{(1)}| - qA_\mu\hat{\gamma}^\mu|\psi\rangle$. If the addition corresponds to an electromagnetic wave of amplitude f and frequency ω, in vanishing gravity, there is a resonance at $\omega = 2E$ that leads to

$$|\beta(t)|^2 = \left(\frac{qkf}{m^2\omega}\right)^2 \cos^2(2Ex_0)\,.$$

If gravity is also present, the resonance condition becomes $\omega = 2E, C = qkf/m^2\omega \equiv A$, with C represented by the terms of (6.63) in curly brackets, and $|\beta(t)|^2 = (A/m)^2 \sin^2(Et)$. The prospects of achieving resonance in laboratory conditions in the near future do not appear favourable.

ZB appears to be universal in condensed matter physics and is the subject of recent, intense research [1]. It is in this area that lie the best opportunities to observe ZB.

Entirely similar conclusions can be reached for the covariant Klein–Gordon equation [44,45] and for all known relativistic wave equations.

In conclusion, the result that static electric and magnetic fields in flat space–time excite a field-dependent ZB can be extended not only to electromagnetic fields of any type in curved space–time, but also to any gravitational fields of weak to intermediate strength and to both fermions and bosons.

Particle propagation is affected by gravitational and electromagnetic Berry phases. They imply gauge structures that mix the field-free states giving rise to oscillations of frequency at least as high as $2m$. This action can be interpreted, in the gravitational case, as an example of how the curvature of space–time can affect Hilbert space by determining transitions between states of positive and negative energy. The transitions involve $\hbar$. Though resonance conditions between ZB and the external fields exist in principle, their realization for particles in vacuum seems unlikely at present. The significance of the results is related to the role played by Berry phase and the related potential $K_\lambda(x, x_0)$ in the mixing of positive and negative energy states that are necessarily contained in the eigenfunctions of relativistic particles. ZB oscillations appear as the particles strive to conserve energy–momentum and angular momentum along their world lines.

Problems

6.1 *Using the method outlined in Sect. 5.1, determine the transition amplitude for a photon that, after travelling through a gravitational field, produces a fermion–antifermion couple.*

6.2 *Consider an electron beam propagating in the plane* $z = 0$ *with momentum* $\mathbf{k} = (k\cos\omega t, k\sin\omega t, 0)$ *and acceleration* $\mathbf{a} = (-\omega^2 x, -\omega^2 y, 0)$. *Find the probability of transition between the spin-up and spin-down components of the beam.*

6.3 ***Problem Heading***

6.1 *It is possible to pass from the process* $f \to f' + \gamma$ *to* $f + \gamma \to f'$ *with the replacement* $P_\mu = p'_\mu - \ell_\mu$. *From* $f + \gamma \to f'$, *one arrives at the process* $\gamma \to f + f'$ *by the replacement* $-P_\mu + p'_\mu = \ell_\mu$ *(substitution law).*

6.2 *From Eq. (6.32), one finds* $P_{2\to 1} = [\frac{k^3}{E}(\Omega R + \frac{k^3}{E+m})\sin(2\Omega t)]^2$, *where* R *is the radius described by the wave packet in the plane* $z = 0$.

References

1. Zawadzki, V., Rusin, T.M.: J. Phys.:Condens. Matter. **23**, 143201 (2011)
2. Cai, Y.Q., Papini, G.: Phys. Rev. Lett. **66**, 1259 (1991); **68**, 3811 (1992)
3. Papini, G.: Advances in the Interplay Between Quantum and Gravity Physics, p. 317. In: Bergmann, P.G., de Sabbata, V. (eds.) Kluwer Academic, Dordrecht (2004). arXiv:gr-qc/0110056
4. Papini, G.: Relativity in Rotating Frames. In: Rizzi, G., Ruggiero, M.L. (eds.) Kluwer Academic, Dordrecht (2004). Ch.16 and references therein, arXiv:gr-qc/0304082

5. Lambiase, G., Papini, G., Punzi, R., Scarpetta, G.: Phys. Rev. D **71**, 073011 (2005)
6. Papini, G.: Gen. Rel. Gravit. **40**, 1117 (2008)
7. Papini, G., Scarpetta, G., Feoli, A., Lambiase, G.: Int. J. Mod. Phys. D **18**, 485 (2009)
8. Jauch, J.M., Rohrlich, F.: The Theory of Photons and Electrons. Springer, New York (1976)
9. See, e.g., Papini, G., Valluri, S.R.: Phys. Rep. **33**, 51 (1977)
10. Misner, C.W., Thorne, K.S., Wheeler, J.A.: Gravitation. Freeman, San Francisco (1973)
11. See, e.g., Renton, P.: Electroweak Interactions. Cambridge University Press, Cambridge (1990)
12. Weinberg, S.: Phys. Rev. B **140**, 516 (1965)
13. Cai, Y.Q., Papini, G.: Class. Quantum Grav. **6**, 407 (1989)
14. Cai, Y.Q., Papini, G.: Class. Quantum Grav. **7**, 269 (1990)
15. Ziese, M., Thornton, M.J. (eds.): Spin Electronics. Springer-Verlag, Berlin (2001)
16. For a comprehensive review see: Igor Žutić, Fabian, J., Sarma, S.D.: Rev. Mod. Phys. **76**, 323 (2004)
17. Bauer, G.E., Bretzel, S., Brataas, A., Tserkovnyak, Y.: Phys. Rev. B **81**, 024427 (2010)
18. De Oliveira, C.G., Tiomno, J.: Nuovo Cimento **24**, 672 (1962)
19. Peres, A.: Suppl. Nuovo Cimento **24**, 389 (1962)
20. Hehl, F.W., Ni, W.-T.: Phys. Rev. D **42**, 2045 (1990)
21. Walter, T.: Grandy. Relativistic Quantum Mechanics of Leptons and Fields. Kluwer Academic Publishers, Dordrecht (1990)
22. Mashhoon, B.: Phys. Rev. Lett. **61**, 2639 (1988); Phys. Lett. A **139**, 103 (1989); **143**, 176 (1990); **145**, 147 (1990); Phys. Rev. Lett. **68**, 3812 (1992)
23. Jackson, J.D.: Classical Electrodynamics. Wiley, Inc. (1999)
24. Matsuo, M., Ieda, J., Saitoh, E., Mackawa, S.: Phys. Rev. Lett. **106**, 076601 (2011)
25. Chowdhury, D., Basu, B.: Physica B **448**, 155 (2014)
26. Kobayashi, D., Yoshikawa, T., Matsuo, M., Iguchi, R., Maekawa, S., Saitoh, E., Nozaki, Y.: Phys. Rev. Lett. **119**, 077202 (2017)
27. Bakke, K.: arXiv:1307.2847v1 [quantum-ph] 10 Jul 2013. Mota, H.F., Bakke, K.: arXiv:1401.378v1 [hep-th] 15 Jan 2014
28. Itzykson, C., Zuber, J.B.: Quantum Field Theory. McGraw-Hill Inc., New York (1980)
29. Papini, G.: Mod. Phys. Lett. A **29**, 1450075 (2014)
30. Papini, G.: Int. J. Mod. Phys. D **26**, 1750137 (2017)
31. Papini, G.: Mod. Phys. Lett. A **35**, 205025 (2020)
32. Berry, M.V.: Singularities in Waves and Rays. Les Houches, Session XXXV, Balian, R., et al. (eds.) 1980-Physique des Défauts/Physics of Defects, North-Holland Publishing Company (1981)
33. Lense, J., Thirring, H.: Z. Phys. **19**, 156 (1918); (English translation: Mashhoon, B., Hehl, F.W., Theiss, D.S.: Gen. Rel. Grav. **16**, 711 (1984))
34. Einstein, A.: Königliche Preussiche Akademie der Wissenschaften (Berlin), Sitzungsberichte, 154–167 (1918)
35. Papini, G.: Phys. Rev. D **82**, 024041 (2010)
36. Nye, J.F., Berry, M.V.: Proc. R. Soc. Lond. A. **336**, 165 (1074); Nye, J.F.: Proc. R. Soc. Lond. A **378**, 219 (1981)
37. Karimi, E., Marrucci, L., Grillo. V., Santamato, E.: Phys. Rev. Lett. **108**, 044801 (2012)
38. Larocque, H., Bouchard, F., Grillo, V., Sit, A., Fabbroni, S., Dunin-Borkowski, R.E., Padgett, M.J., Boyd, R.W., Karimi, E.: Phys. Rev. Lett. **117**, 154801 (2016)
39. Harris, J., Grillo, V., Mafakheri, E., Gazzadi, G.C., Fabbroni, S., Boyd, R.W., Karimi, E.: Nat. Phys. **11**, 629 (2015)
40. Bialynicki-Birula, I., Bialynicka-Birula, Z.: Phys. Rev. Lett. **118**, 114801 (2017)
41. Barnett, S.M.: Phys. Rev. Lett. **118**, 114802 (2017)
42. Matsumoto, H., Umezawa, H.: Phys. Rev. D **14**, 3536 (1976)
43. Leplae, L., Mancini, F., Umezawa, H.: Phys. Rep. **10**, 151 (1974)
44. Papini, G.: Phys. Lett. A **376**, 1287 (2012)
45. Feshbach, H., Villars, F.: Rev. Mod. Phys. **30**, 24 (1958)

Other Developments

7

Abstract

Advances in instrumentation place more stringent limits on violations of conservation laws. This is the case for discrete symmetries where spin plays a fundamental role. Spin and gravitation endow space–time with a measure of chirality. Axions introduce chirality in the universe and are considered leading candidates for dark matter. Their discovery would provide a solution to two of the major problems of modern physics, the violation of strong CP invariance and the origin of dark matter. The searches for axions and for their interactions with spin and matter are the subjects of this chapter. These developments require the formulation of appropriate forms of the Bargmann–Michel–Telegdi equation [1,2] and of a theory of axion electrodynamics [3,4]. In a purely geometrical context, chirality could also be represented by torsion. Experimental searches for some of these effects are under way [5].

7.1 Scalar–Pseudoscalar Coupling and the Search for Axions

Spin-0 fields appear in several branches of theoretical particle physics, such as the standard model and string theory and in all theories concerned with spontaneous symmetry breaking. They also appear in several versions of scalar-tensor theories of gravitation. Some of these theories [6–9] contemplate the coupling of gravitation to spin and the presence of gravitational dipole moments. Some other approaches [10,11] consider monopole–dipole potentials generated by the exchange of a light pseudoscalar field. Constraints have been established on the gravitational dipole moment of electrons and neutrons [12] and protons [13]. These measurements find that the spin-dependent component of the gravitational energy of the proton in the field of Earth is smaller than 3×10^{-8} eV, a three orders of magnitude improvement over laboratory constraints on long-range monopole–dipole interactions.

Spin-0 fields figure prominently in major physics problems like dark matter and matter–antimatter asymmetry in the universe. Axion-like particles introduced by

G. Lambiase and G. Papini, *The Interaction of Spin with Gravity in Particle Physics*, Lecture Notes in Physics 993, https://doi.org/10.1007/978-3-030-84771-5_7

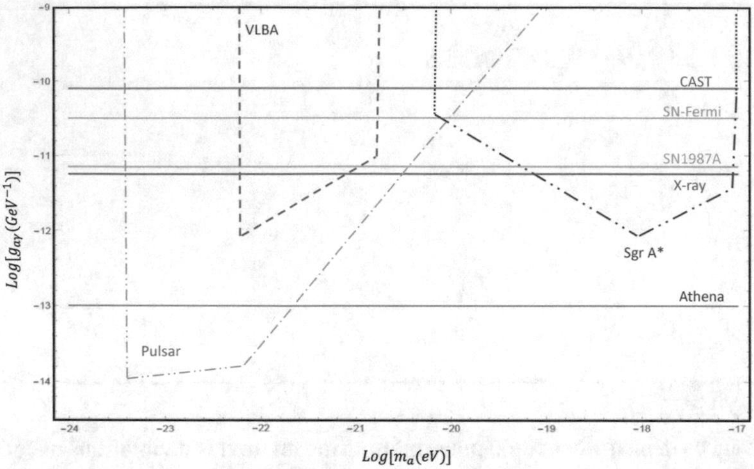

Fig. 7.1 Upper limits of the ALP–photon coupling $g_{a\gamma}$ for different ALPs mass m_a. The ranges $10^{-14} \lesssim g_{a\gamma}(\mathrm{GeV}^{-1}) \lesssim 10^{-10}$ and $10^{-24} \lesssim m_a(\mathrm{eV}) \lesssim 10^{-17}$ follow from the observations of the CAST experiment [25], the Fermi-LAT observation of a population of SNe [26], the SN1987A [27], the X-ray observation of star clusters [28], the VLBA polarization observations of jets from active galaxies [29], pulsars [30], and Sgr A* [31]

Peccei and Quinn [14, 15] in the guise of scalar–pseudoscalar fields in order to solve the problem of strong CP invariance would violate parity and time-reversal symmetries and also explain the matter–antimatter asymmetry mentioned. Nearly massless pseudoscalar fields could give rise to long-range interactions with spin and therefore be of interest in astrophysics and cosmology. It has, in particular, been shown that in a scalar-tensor theory of gravitation there would be coupling of spin to gravitation and particles would acquire a gravitational dipole moment, or an anomalous gravito-magnetic moment, or gravito-electric dipole moment. Axions are considered as attractive candidates for dark matter.

Spin-axion coupling is particularly important. It is in fact hoped that the effect that axions are predicted to have on the spin of particles in storage rings, or in the vicinity of strong gravitational and axion fields (axion stars) would be observable, thus leading to the observation of axions and their properties.

The interaction of an axion (pseudoscalar field) on the spin of massive particles is considered within the frame of a covariant model [2]. Covariance is required by the relativistic motion of the particles involved under the action of strong fields. The model studies the evolution of spin by including axionic modifications in the Bargmann–Michel–Telegdi equation [1, 2] in an environment of pseudoscalar, gravitational, and electromagnetic fields. The result of this generalization is that axion fields can generate a spin precession. The pseudoscalar field renders space–time chiral and the left- and right-hand rotations of the spin are not equivalent. A short introduction to axion electrodynamics is given in the next section.

Studies of axions in different contexts have been provided in Refs. [16–24]. Particularly important is the role of axion-like particles (ALPs), which are among the favourite candidates as Dark Matter particles (see Fig. 7.1).

7.2 Axion Electrodynamics

Axion electrodynamics [3] is based on the action functional

$$L = -\frac{1}{4}F_{\mu\mu}F^{\mu\nu} + \alpha F_{\mu\nu}F^{\star\mu\nu}\phi + \frac{1}{2}\partial_\mu\phi\partial^\mu\phi - \frac{1}{2}m_a^2 , \tag{7.1}$$

where $F_{\alpha\beta} = A_{\beta,\alpha} - A_{\alpha,\beta}$ is the electromagnetic field tensor and $F^{\star\mu\nu} = \frac{1}{2}\epsilon^{\mu\nu\alpha\beta}F_{\alpha\beta}$ is its dual. The constant α is model dependent and will be determined below for the particular case of axions interacting with superconductors. The classical field equations that can be derived from (7.1) are

$$E^i_{,i} = \alpha B^i \phi_{,i} , \tag{7.2}$$

$$\epsilon^{ijk}B_{j,k} - E^i_{,0} = \alpha\left(\epsilon^{ijk}E_j\phi_{,k} - B^i\phi_{,0}\right) , \tag{7.3}$$

$$\partial_\mu\partial^\mu\phi = \alpha E_i B^i - m_a^2\phi . \tag{7.4}$$

Because of the introduction of the potentials A_μ, the relationship $F^{*\mu\nu}_{,\nu} = 0$ is identically satisfied and the conservation equation $(F^{\mu\nu} + \phi F^{\star\mu\nu})_{,\mu} = 0$ simplifies to

$$\partial_\nu F^{\mu\nu} = -F^{\star\mu\nu}\phi_{,\nu} . \tag{7.5}$$

The cross section for the axion–photon conversion in strong inhomogeneous magnetic fields B_0 can be calculated from Eqs. (7.1) and (7.2) and is

$$\sigma = \frac{e^2}{16 v_a \pi^2}\sum_\lambda \int d^3k_\gamma \delta(E_\gamma - E_a)|\int_V d^3x e^{i(\mathbf{k}_\gamma - \mathbf{k}_a)\cdot\mathbf{x}}\mathbf{B_0}(\mathbf{x}) \cdot \boldsymbol{\epsilon}(\mathbf{k}_{\gamma,\lambda}|^2 \tag{7.6}$$

where the sum is over the photon polarizations. A detection plan can now be devised: from σ one calculates the flux of photons, produced at the detector by the conversion of axions into photons, and adjust the detector parameters according to the axion spectrum. The experiments proposed apply to astrophysical sources of axions such as the Sun and the halo of our galaxy.

7.3 The Extended Bargmann–Michel–Telegdi Model

When considering charged particles with spin in electromagnetic and axionic fields, it is useful to consider generalizations of the Bargmann–Michel–Telegdi (BMT) equation [1,2]. Their derivation starts from the evolution of the particle momentum and spin, represented by the equations

$$\frac{Dp^{\mu}}{D\tau} = F^{\mu} = \frac{e}{m} F^{\mu}_{\ \nu} p^{\nu}\,, \tag{7.7}$$

where F^{μ} represents the Lorentz force, and the spin vector S^{μ} satisfies the equation

$$\frac{DS^{\mu}}{D\tau} = G^{\mu}\,. \tag{7.8}$$

If the particle mass is constant, then it follows from (7.7) that $p^{\alpha}F_{\alpha} = 0$. If, in addition, $S_{\alpha}S^{\alpha}$ is also constant, then it follows from (7.8) that $S_{\alpha}G^{\alpha} = 0$. If one finally assumes that $S_{\alpha}p^{\alpha} = 0$, then $F_{\alpha}S^{\alpha} + G_{\alpha}p^{\alpha} = 0$. The last three equations are satisfied if an antisymmetric tensor $\omega_{\alpha\beta}$ exists for which $F^{\alpha} = \omega_{\alpha\beta}p^{\beta}$ and $G^{\alpha} = \omega^{\alpha\beta}S_{\beta}$. In the BMT model,

$$\omega^{\alpha\beta} = \frac{e}{2m}\left(gF^{\alpha\beta} + \frac{g-2}{m^2}(\delta^{\alpha}_{\sigma}\delta^{\beta}_{\tau} - \delta^{\alpha}_{\tau}\delta^{\beta}_{\sigma})p_{\rho}F^{\rho\sigma}p^{\tau} \right). \tag{7.9}$$

Minimal contributions to $\omega_{\alpha\beta}$ linear in ϕ can be obtained from (7.9) by replacing the charge e with the pseudoscalar charge $e\phi$ and $F_{\mu\nu}$ with its dual and by letting $g \to g_A$. Similarly, minimal contributions in $\phi_{,\mu}$ can be obtained from (7.9) by the replacements $e \to e\mu p^{\alpha}\phi_{,\alpha}$, $g \to g_G$, $F^{\alpha\beta} \to F^{\star\alpha\beta}$, where μ is a constant parameter. The flat space–time generalization of the BMT equation in the presence of axions is therefore [2]

$$\begin{aligned} \frac{DS^{\mu}}{D\tau} = \frac{e}{2m}&\left[F^{\mu\nu}S_{\nu} + \frac{(g-2)}{m^2} p^{\mu}F_{\alpha\beta}S^{\alpha}p^{\beta} \right] \\ &+\frac{e\lambda}{2m}\phi\left[g_A F^{\star\mu\nu}S_{\nu} + \frac{g_A-2}{m^2}p^{\mu}F^{\star}_{\alpha\beta}S^{\alpha}p^{\beta} \right] \\ &+\frac{e\mu}{2m}(p^{\alpha}\phi_{,\alpha})\left[g_G F^{\star\mu\nu}S_{\nu} + \frac{g_G-2}{m^2}p^{\mu}F^{\star}_{\alpha\beta}S^{\alpha}p^{\beta} \right] \\ &+\frac{e\mu}{2m}(p^{\alpha}\phi_{,\alpha})\omega_{23}\epsilon^{\mu\alpha\beta\gamma}S_{\alpha}p_{\gamma}p^{\delta}F_{\beta\delta} \\ &+\frac{\omega_{24}}{m}\phi^{,\rho}\epsilon^{\mu\nu\sigma\tau}S_{\nu}p_{\tau}F_{\sigma\rho} + \frac{\omega_{25}}{m}\phi_{,\alpha}\epsilon^{\mu\sigma\alpha\tau}S_{\sigma}p_{\tau}\,, \end{aligned} \tag{7.10}$$

and λ is also a constant parameter. The first term on the r.h.s. of (7.10) is the usual BMT term for a particle with $g \neq 2$. The other terms are axion-induced terms and $\omega_{23}, \omega_{24}, \omega_{25}$ are coupling constants. The term with ω_{25} does not contain $F_{\mu\nu}$ and therefore represents a direct spin–axion coupling. With the help of (7.10) strategies can be designed to observe, by means of polarized beams of particles in storage rings, the spin precession induced by axion stars, relic dark matter axions, axions distributed around spherically symmetric static objects, and axions in a gravitational-wave field. The pseudoscalar axion field confers some chirality to space–time such that left-handed and right-handed rotations of a particle's spin are not equivalent. The methods discussed above have lead to the proposal of several axion searches [5].

7.4 Space–Times with Torsion

An interesting idea born out of attempts to conciliate general relativity with quantum theory is the representation of spin as the source of a field. In the Einstein–Cartan–Sciama–Kibble theory [6], this requires the generalization of the usual Riemannian space–time V_4 to the manifold U_4. In this framework, the spin of a particle is related to the torsion of space–time, very much like mass is responsible for curvature.

The idea of torsion arises in any fundamental theory. For instance, a torsion field appears in (super)string theory [32], in four and in higher dimensional theories as in the theory of Kaluza–Klein [33], in theories of gravity formulated in terms of twistors [34], in supergravity [35] and in cosmology [6,36–40].

Some consequences of a torsion field coupled to particles, neutrinos, in particular, are discussed below.

7.4.1 Spin-Flip Transitions in Space–Times with Torsion

It is shown in this section that the helicity of a fermion is not conserved in a space with torsion. This is found by computing the time variation of the helicity operator in the Heisenberg representation and by showing that it does not vanish.

Writing the Dirac equation in curved space–time, using the vierbein formalism [41], one finds that the connection in non-holonomic coordinates is given by [42]

$$\Gamma_{abc} = -\Omega_{abc} + \Omega_{bca} - \Omega_{cab} + S_{abc}\,, \tag{7.11}$$

where $\Omega^c_{\alpha\beta} \equiv e^c_{[\alpha,\beta]}$, $\Omega^c_{ab} = e^\alpha_a\, e^\beta_b\, c^c_\sigma\, \Omega^\sigma_{\alpha\beta}$, and S_{abc} is the antisymmetric part of the affine connection. Reverting to ordinary units, the covariant derivative and the Dirac equation are given by $D_\mu \equiv \partial_\mu - \frac{1}{4}\Gamma_{\mu ab}\gamma^a\gamma^b$, and

$$\gamma^\alpha \psi_{,\alpha} + i\frac{mc}{\hbar}\psi = \frac{1}{4}S_{\alpha\beta\sigma}\gamma^\alpha\gamma^\beta\gamma^\sigma,\psi \tag{7.12}$$

respectively. To emphasize the effects due to torsion, the gravitational effects on the spin flip are neglected, which is equivalent to neglecting the Ω_{abc} terms in Eq. (7.11). The corresponding Hamiltonian operator is

$$H = c\boldsymbol{\alpha}\cdot\mathbf{p} + mc^2\beta + \frac{i}{4}S_{\alpha\beta\sigma}\gamma^0\gamma^\alpha\gamma^\beta\gamma^\sigma \equiv H_0 + H'\,, \tag{7.13}$$

where H' denotes the perturbation to the unperturbed Hamiltonian $H_0 = c\boldsymbol{\alpha}\cdot\mathbf{p} + mc^2\beta$.

The helicity operator is defined as [43] $h = \mathbf{\Sigma} \cdot \hat{\mathbf{p}}$ where $\mathbf{\Sigma} = \sigma^i \bigotimes I_{2\times 2}$, $I_{2\times 2}$ is the 2×2 unit matrix, while $\hat{p}^i = \frac{p^i}{|\mathbf{p}|}$. Here σ^i, $i = 1, 2, 3$ are the Pauli matrices and $p^\mu = (p^0, \mathbf{p})$ is the momentum. In the Heisenberg representation, the evolution of the helicity operator is $i\hbar\dot{\hat{h}} = [\hat{h}, H]$, and using (7.13) one gets

$$i\dot{h} = \frac{cp^k}{4|\mathbf{p}|}\varepsilon_{ijk} S_{\alpha\beta\sigma}\gamma^0 \left[g^{i\sigma}\gamma^\alpha\gamma^\beta\gamma^j + 2g^{i\alpha}g^{j\beta}\gamma^\sigma\right], \tag{7.14}$$

which implies that $\dot{h} \neq 0$, i.e. the helicity of a fermion particle is not conserved.

One can also calculate the probability of the helicity flip induced by the torsion term in Eq. (7.13) [44]. To this aim one considers the totally antisymmetric dual or a null vector, $S^\sigma = (|\mathbf{S}|, \mathbf{S})$, the approximation $g_{\mu\nu} \sim \eta_{\mu\nu}$, and re-casts the Hamiltonian (7.13) in the form

$$H' = -\frac{3\hbar c|\mathbf{S}|}{2}\gamma^5 + i\frac{3}{2}\mathbf{S} \cdot \begin{pmatrix} \sigma & 0 \\ 0 & \sigma \end{pmatrix}. \tag{7.15}$$

Writing the spinor as $|\psi >= \begin{pmatrix} |\psi_R > \\ |\psi_L > \end{pmatrix}$, where the states $|\psi_R >$ and $|\psi_L >$ are eigenstates of energy, $H_0|\psi_{R/L} >= E|\psi_{R/L} >$, one finds that the evolution of the state $|\psi(t) >$ at time t can be written as [44]

$$|\psi(t) > e^{-i(E/\hbar + 3c|\mathbf{S}|/4)t}\, e^{-3c|\mathbf{S}|^2 t}\, [\cos\frac{c|S|}{2}t\, |\psi_R > + \sin\frac{c|S|}{2}t\, |\psi_L >] .$$

The probability to find the particle in state $|\psi_R >$ at time t is $P_R(t) \sim \cos^2(3c|\mathbf{S}|/4)t$, while the probability that the spin flip occurs is $P_L(t) \sim \sin^2(3c|\mathbf{S}/4)t$. The corresponding spin-flip frequency is $\omega = 3c|\mathbf{S}|/4$, while the characteristic length is $L = 8\pi/3|\mathbf{S}|$.

The free fall universality measurements of Ref. [45] place an upper limit of $5.4 \times 10^{-6} m^{-2}$ on a possible gradient field of space–time torsion.

7.4.2 More About Torsion

As mentioned at the beginning of this chapter, the possibility to use torsion to study spin has raised considerable interest in this field. At stake are here the notions that spin creates a field that affects other particles by means of forces and torques [8]. It seems natural, in fact, to extend general relativity, that is, a geometrical theory of gravitation, by introducing spin as a geometrical quantity to represent a quintessential quantum object like spin, that is, quantized in units of $\hbar/2$.

Investigations of the behaviour of spin$-1/2$ particles interacting with torsion have been given by Shapiro [7] and developments pertaining to the Dirac equation in curved space–times with torsion have been carried out by Audretsch [46], while some physical effects have been discussed by Lämmerzahl [47].

The derivation of a torsion generated Berry phase is particularly relevant to the topics discussed in this work. The general structure of this phase can be surmised from (7.15) where the vector **S** interacts with spin as a magnetic field does with the magnetic moment of spin$-1/2$ particles. It is therefore useful to introduce the polar and azimuthal angles that characterize the unit vector $\tilde{\mathbf{n}}$ and consider the eigenvectors of $\boldsymbol{\sigma}\cdot\hat{\mathbf{n}}\chi=\chi$. They are [48]

$$\chi_+=\begin{pmatrix}\cos\frac{\theta}{2}e^{-i\varphi}\\ \sin\frac{\theta}{2}\end{pmatrix};\;\chi_-=\begin{pmatrix}\sin\frac{\theta}{2}e^{-i\varphi}\\ -\cos\frac{\theta}{2}\end{pmatrix}. \tag{7.16}$$

Writing $S_i=\epsilon_{ijk}\tilde{A_{jk}}$, one obtains the Berry connection [49]

$$\tilde{A}_\varphi=\frac{3S}{2}(\chi_-|i\partial_\varphi\chi_-)=\frac{3S}{2}\sin^2\frac{\theta}{2};\;\tilde{A}_\theta=\frac{3S}{2}(\chi_-|i\partial_\theta\chi_-)=0\,, \tag{7.17}$$

and also the Berry curvature

$$\Pi_{\theta\varphi}=\partial_\theta\tilde{A}_\varphi-\partial_\varphi A_\theta=\frac{3S}{2}\sin\theta\,. \tag{7.18}$$

The contribution of torsion to the Berry phase over a closed contour $\mathcal{C}$ is therefore contained in

$$\oint\Pi_{\theta\phi}d\Omega \tag{7.19}$$

which can be measured, in principle, by interferometric means.

Reference [50] considers a neutral spin-$1/2$ particle endowed with permanent electric and magnetic dipole moments, in the presence of torsion, when the curvature is represented by a cosmic dislocation and the curvature has a conical singularity. The contribution of torsion to the phase of the particle is similar.

7.5 Axions and Berry Phase

Experiments aimed at detecting axions by means of superconductors are at present pursued [51–54]. It is therefore of interest to calculate the phase of Berry using the theory developed in Sect. 1.3.2.

As seen in Sect. 7.2, the axion electrodynamics of Ni and Sikivie is based on Eqs. (7.2)–(7.5) and the fundamental equations

$$(F^{\alpha\beta}+\alpha\phi F^{*\alpha\beta})_{,\beta}=0\,, \tag{7.20}$$

where ϕ is the axion field and α is a still undetermined coupling constant. The electromagnetic field $F_{\mu\nu}=A_{\nu,\mu}-A_{\mu,\nu}$ and its dual $F^{*\mu\nu}$ satisfy Maxwell equations

$$F^{\mu\nu}{}_{,\nu}=-4\pi j^\nu\,, \tag{7.21}$$
$$F^{*\mu\nu}{}_{,\nu}=0\,. \tag{7.22}$$

The quantity $q\phi$ plays the role of an effective pseudo-charge. It follows from (7.20) that

$$F^{\mu\nu}{}_{,\nu} = -\alpha\, \phi_{,\nu}\, F^{*\,\mu\nu}$$

behaves like a current which yields

$$-\,\alpha F^{*}_{0\nu}\phi_{,}{}^{\nu} = \alpha(\nabla\phi \cdot \mathbf{B}) \tag{7.23}$$

and

$$-\,\alpha F^{*}_{i\nu}\phi_{,}{}^{\nu} = -\alpha\left[\frac{\partial\phi}{\partial t}B_i - (\mathbf{E}\times\nabla\phi)_i\right]. \tag{7.24}$$

Equations (7.23) and (7.24) coincide with the first two equations of (7.2). The superconductor is assumed to be adequately described by a gas of scalar particles of charge $q = 2e$ moving against a positively charged background represented by the lattice. Substituting $\psi = \rho(\mathbf{r})e^{i\theta(\mathbf{r})}$ in the expression for the current of a superconductor

$$4\pi\mathbf{j} = \frac{q}{2m}\left(\psi^*\nabla\psi - \psi\nabla\psi^*\right) - \frac{q^2}{m}\psi^*\mathbf{A}\psi\,, \tag{7.25}$$

one obtains

$$4\pi\mathbf{j} = \frac{1}{m}\,(\nabla\theta - q\mathbf{A})\,\rho\,. \tag{7.26}$$

Since inside a superconductor the charge density of electrons is uniform, $\nabla\theta \simeq 0$, one finds

$$\mathbf{j} = -\frac{\rho q}{4\pi m}\mathbf{A}\,. \tag{7.27}$$

This equation, originally proposed by Heinz and Fritz London to explain superconductivity [55], leads immediately to a phase $\chi_\alpha = qA_\alpha$, and to the Meissner effect. In fact, from the equations

$$4\pi\mathbf{j} = \nabla\times\mathbf{B}, \quad \mathbf{j} = -\frac{\rho q}{4\pi m}\mathbf{A}\,,$$

one obtains

$$\nabla\times\nabla\times\mathbf{B} = 4\pi\nabla\times\mathbf{j} = -\nabla^2\mathbf{B} = -\frac{\rho q}{m}\mathbf{B}$$

which describes the Meissner effect. Equation (7.27) can be used to give a value to the constant α introduced in Sect. 7.2. One finds $\alpha = (\rho q/4\pi m)^{-1} = \lambda^2$. It therefore seems reasonable to introduce a phase

$$\chi_{,\alpha} = qA_\alpha - \alpha\phi_{,\nu}F^{*}_{\alpha}{}^{\nu}\,, \tag{7.28}$$

as the appropriate generalization of $\theta_{,\alpha} = qA_\alpha$. This equation also implies that the generalized momentum of the superelectrons is

$$p_\alpha = k_\alpha + qA_\alpha - \alpha\phi_{,\nu}F^{*}_{\alpha}{}^{\nu}\,.$$

Setting

$$\Phi(x) = e^{-i\chi}\phi_0(x)\,, \tag{7.29}$$

where ϕ_0 satisfies the equation

$$(\partial_\mu \partial^\mu + m^2)\phi_0 = 0\,. \tag{7.30}$$

Using (7.28) and (7.30), one gets

$$\begin{aligned}\Delta_\mu \Delta^\mu \Phi + m^2\Phi &= e^{-i\chi}(\partial_\mu - i\chi_{.\mu})(\partial^\mu - i\chi_{,}{}^{\mu})\phi_0 + m^2\, e^{-i\chi}\phi_0 \\ &= e^{-i\chi}\left[(\partial_\mu\partial^\mu + m^2)\phi_0 - i\chi_{,\mu}^{\;\mu} - 2i\chi_{,\mu}\,\phi_{0,}{}^{\mu} - \chi_{,\mu}\chi_{,}{}^{\mu}\phi_0\right] \\ &= e^{-i\chi}\left[-i\chi_{,\mu}^{\;\mu} - 2i\chi_{,\mu}\,\phi_{0,}^{\;\mu} - \chi_{,\mu}\chi_{,}{}^{\mu}\phi_0\right],\end{aligned} \tag{7.31}$$

where $\Delta_\mu = \partial_\mu + i\chi_{,\mu}$. From (7.28), one obtains

$$\chi_{,\mu}^{\;\mu} = qA_{\alpha,}{}^{\alpha} - \alpha\phi_{,\nu\alpha}F^{*\alpha\nu} + \alpha\phi_{,\nu}F^{*\,\nu\alpha}_{\quad\alpha} = 0\,. \tag{7.32}$$

This result follows because the first term vanishes by choosing an electromagnetic gauge such that $\partial^\alpha A_\alpha = 0$, the second term vanishes because two symmetric indices are contracted with two antisymmetric indices (assuming that ϕ is regular), and finally the last term vanishes owing to the fields equations. Inserting (7.32) into (7.31), one obtains

$$(\Delta_\mu \Delta^\mu \Phi + m^2)\Phi = e^{-i\chi}\chi_{,\mu}\left[-2i\phi_{0,}^{\;\mu} - \chi_{,}{}^{\mu}\phi_0\right]. \tag{7.33}$$

The Berry connection is gauge dependent and a gauge transformation can be chosen to dispose of the right-hand side of (7.33). For a plane-wave solution $\phi_0 = e^{ik_\mu x^\mu}$, one finds

$$(\Delta_\mu \Delta^\mu \Phi + m^2)\Phi = e^{-i\chi}\chi_{,\mu}\left[2k^\mu - \chi_{,}{}^{\mu}\right]\phi_0\,,$$

which yields the gauge term that makes the right-hand side of (7.33) vanish. Therefore, the effect of the axion on a scalar particle with charge (here a relativistic super-electron) can be contained in a phase as in $\Phi = e^{-i\chi}\phi_0$, with

$$\chi = \int^x dz^\lambda (qA_\lambda - \alpha\phi_{,\nu}F^{*\;\nu}_{\lambda})\,, \tag{7.34}$$

which is the Berry phase induced by the axions. For a closed path linking a multiply connected region of the superconductor, one also finds

$$\chi = \oint dz^\lambda (qA_\lambda - \alpha\phi_{,\nu}F^{*\;\nu}_{\lambda}) = 2n\pi\,, \tag{7.35}$$

as required in order to make the wave function of the superelectrons single valued. Equation (7.35) is gauge invariant and observable, in principle. Quantization of the flux then occurs for the paths encircling the topological singularity

$$q \oint A_\alpha dz^\alpha = q \int_\Sigma \mathbf{B} \cdot d\mathbf{S} = -\alpha \oint dz^\lambda \phi_{,\nu} F^{*\ \nu}_{\lambda} . \tag{7.36}$$

It follows from (7.36) that the current $\phi_{,\alpha} F^{*\ \nu}_{\alpha}$ generated by an axion produces a magnetic field in a loop of wire, hence a current, that could, in principle, be observed.

If, in particular, $n = 0$, then

$$\oint dz^\alpha q A_\alpha = \alpha \oint \phi_{,\nu} F^{*\nu}_{\alpha} dz^\alpha . \tag{7.37}$$

The explicit form of (7.37) can be calculated using for $F^{*\mu\nu}$ the matrix

$$F^{*\mu\nu} = \begin{pmatrix} 0 & -B_x & -B_y & -B_z \\ B_x & 0 & E_z & -E_y \\ B_y & -E_z & 0 & E_x \\ B_z & E_y & -E_x & 0 \end{pmatrix} , \tag{7.38}$$

which, using (7.23) and (7.24), leads to

$$\oint dz^\alpha A_\alpha = \frac{\alpha}{q} \oint \left[dz^0(-\mathbf{B} \cdot \nabla\phi) + d\mathbf{z} \cdot \mathbf{B} \frac{\partial \phi}{\partial t} - d\mathbf{z} \cdot (\mathbf{E} \times \nabla\phi) \right] = \Sigma_1 + \Sigma_2 + \Sigma_3 ,$$

where A_α refers to the electromagnetic field in the superconductor, **B** and **E** are generated by ϕ. Σ_1 and Σ_2 have opposite signs and tend to compensate each other.

Detection schemes used in current laboratory experiments concentrate on the term Σ_2. They are based on Josephson junctions that are small devices in which an axion flux generates an oscillating current, with a frequency given by the axion mass. The dimensions of the devices, small relative to the Compton wave length of an axion in the range of dark matter axions from μeV to meV, render negligible the gradient terms Σ_1, Σ_3. Introducing the axion misalignment parameter $\vartheta = \phi/f$, where f is the axion coupling constant, the third equation (7.2) becomes [56]

$$\frac{\partial^2 \vartheta}{\partial t^2} + \Gamma \frac{\partial \vartheta}{\partial t} - \nabla^2 \vartheta + m^2 \theta \sin \vartheta = -\delta \mathbf{E} \cdot \mathbf{B} , \tag{7.39}$$

where Γ is a small damping constant and δ a model-dependent constant. The axion-induced current density is therefore $\mathbf{j} = -\beta \mathbf{B} \frac{\partial \vartheta}{\partial t}$, which gives rise to the current $\mathbf{I} = A\mathbf{j}$, where A is a small area of the junction's weak link where $B \neq 0$ and β also is a model-dependent constant. The current **I**, which for dark matter axions is oscillating, gives rise to the well-known AC Josephson effect and the device, if operated at resonance, could yield sizable gains.

An estimate of the current due to the term Σ_2 can be obtained by assuming that B remains approximately constant along the integration path. Then the current I corresponding to Σ_2 becomes

$$I = -\frac{\alpha}{\mathcal{L}q}\int dt\frac{d\mathbf{z}}{dt}\cdot\mathbf{B}\frac{\partial\phi}{\partial t} \approx -\frac{4\pi\lambda^2}{\mathcal{L}q}(\mathbf{v}_0\cdot\mathbf{B}), \vartheta \tag{7.40}$$

where $\mathbf{v_0}$ is the velocity of the superelectrons and $\mathcal{L}$ the self-inductance of the superconducting circuit. A direct measurement of Berry's phase χ, through its associated current (7.40), would be somewhat independent of axion mass and resonance conditions and represent a sort of particle analogue of a "wide band" detector.

In stationary conditions, the flow of axions also induces an electric field in the superconductor. In the absence of axions, one must have inside the superconductor $\mathbf{F} = -\nabla V = 0$, hence $\mathbf{E} = 0$. From the potential

$$V = -qA_0 - \frac{fq\rho}{m}(\nabla\phi\cdot\mathbf{B}) \tag{7.41}$$

inside the superconductor, one must now have an electric field

$$\mathbf{E} = \frac{f\rho}{m}\nabla(\nabla\phi\cdot\mathbf{B})\ . \tag{7.42}$$

These results are in line with those of Sects. 1.3.1, 1.3.2, and 3.1.1.

Notice that the contribution of the axion field in (7.34) does not depend on the mass of superelectrons and remains therefore unchanged for particles of any mass. If the particles circulating in the interferometer are neutral, then $q = 0$ and the first term in (7.34) vanishes. If the particles circulating in the interferometer are not superfluid and $j_{\mu}\neq\alpha\chi_{,\mu}$, axions can still make themselves felt through the dispersion relations $p_\alpha p^\alpha$. These affect in turn the resonance conditions of wave guides and cavities through the boundary conditions. For rectangular wave guides of sizes a and b, for instance, the cutoff frequency becomes to $\mathcal{O}\left(\frac{\partial\phi}{\partial t}\right)$ for simplicity,

$$\omega_{mn} = \pi\left[\left(\frac{m^2}{a^2}+\frac{n^2}{b^2}\right) + 2\alpha k^\alpha\tilde{\omega}_\nu F^{*\ \nu}_{\alpha}\right]^{\frac{1}{2}} \tag{7.43}$$

where m and n are positive integers. Because $\alpha \sim a^2$, large wave guides, or cavities, are in this case favoured in experiments aimed at detecting axion-induced Berry phases.

Problems

7.1 Estimate $\frac{\partial\phi}{\partial x}$ and calculate Σ_1.

7.2 Estimate the electric field E in the superconductor and calculate Σ_3.

7.3 Calculate the magnetic field and current generated by Σ_1 in a circuit assuming that the self-inductance of the circuit is L.

7.4 Problem Heading

7.1 Assume that $\nabla\phi$ and B are approximately homogeneous in a superconducting strip of volume $w\ell\lambda$, where λ is the penetration depth of the magnetic field. Then $\nabla\phi \sim m_\phi/\lambda$. Also $z_0 \sim \ell/\beta_0$ where the charge carrier speed $\beta_0 \sim 10^{-5}$. Take $B \sim B_c \sim 10^{-2}T$, where B_c is the critical field for many superconductors. The result is

$$\Sigma_1 \sim \frac{-\alpha B_c m_\phi}{q\lambda}\frac{\ell}{\beta_0}. \tag{7.44}$$

7.2 Start from the equations of motion of an electron under the action of electric and magnetic fields

$$\frac{d\mathbf{v}}{dt} = \frac{q}{m}\left[\mathbf{E} + (\boldsymbol{\beta} \times \mathbf{B})\right]. \tag{7.45}$$

By introducing the local acceleration $\partial\mathbf{v}/\partial t$, the equations of motion become

$$\frac{\partial\mathbf{w}}{\partial t} = \nabla \times (\mathbf{v} \times \mathbf{w}). \tag{7.46}$$

Show that when $\mathbf{w} = 0$ (non-viscous electronic fluid) and $\partial\mathbf{v}/\partial t \approx 0$, the electric field in the superconductor is $\mathbf{E} \approx (m/2q)\nabla(v^2)$. Assume that $\mathbf{v}$ at the surface is smaller than the value allowed by $\mathbf{B}_c$ and substitute the value found in Σ_3.
7.3 Proceed as in 6.1 to calculate Σ_1. Show that the current in a circuit is $I = \Sigma_1/qL$.

References

1. Bargmann, V., Michel, L., Telegdi, V.: Phys. Rev. Lett. **2**, 435 (1959)
2. Balakin, A.B., Popov, V.A.: Phys. Rev. D **92**, 105025 (2015)
3. Ni, W.-T.: Phys. Rev. Lett. **38**, 301 (1977)
4. Sikivie, P.: Phys. Rev. Lett. **51**, 1415 (1983)
5. See, e.g., Chang, S.P., Haciomeruglu, S., Soohyung Lee, O.K., Park, S., Semertzidis, Y.K.: XVII International Workshop on Polarizes sources, Targets and Polarimetry, Kaist, South Korea, 16–20 Oct 2017
6. Hehl, F.W., von der Heyde, P., Kerlick, G.D., Nester, J.M.: Rev. Mod. Phys. **48**, 393 (1976)
7. Shapiro, I.L.: Phys. Rep. **357**, 113 (2002)
8. Hammond, T.T.: Rep. Prog. Phys. **65**, 599 (2002)
9. Kostelecký, V.A., Russell, N., Tasson, J.D.: Phys. Rev. Lett. **100**, 111102 (2008)
10. Moody, J.E., Wilczek, F.: Phys. Rev. D **30**, 130 (1984)
11. Flambaum, V., Lambert, S., Pospelov, M.: Phys. Rev. D **80**, 105021 (2009)
12. Jackson Kimball, D.F.: New J. Phys. **17**, 073008 (2015)
13. Jackson Kimball, D.F., Dudley, J., Li, Y., Patel, D., Valdez, J.: Phys. Rev. D **96**, 075004 (2017)
14. Peccei, R.D., Quinn, H.R.: Phys. Rev. Lett. **38**, 1440 (1977)
15. Weinberg, S.: Phys. Rev. Lett. **40**, 223 (1978); Wilczek, F.: Phys. Rev. Lett. **40**, 279 (1978)

16. Lambiase, G., Mohanty, S.: Mon. Not. R. Astron. Soc. **494**, 5961 (2020)
17. Capolupo, A., Lambiase, G., Quaranta, A., Giampaolo, S.M.: Phys. Lett. B **804**, 135407 (2020)
18. Capolupo, A., De Martino, I., Lambiase, G., Stabile, A.: Phys. Lett. B **790**, 427 (2019)
19. Marsh, D.J.E.: Phys. Rep. **643**, 1 (2016)
20. Graham, P.W., Rajendran, S.: Phys. Rev. D **84**, 055013 (2011)
21. Raffelt, G.G.: Stars as Laboratories for Fundamental Physics: The Astrophysics of Neutrinos, Axions, and Other Weakly Interacting Particles. The University Chicago Press (1996)
22. Sikivie, P.: Rev. Mod. Phys. **93**, 015004 (2021)
23. Dror, J.A., Murayama, H., Rodd, N.L.: Phys. Rev. **103**, 115004 (2021)
24. Choi, K., Im, S.H., Shin, C.S.: Annu. Rev. Nucl. Part. Sci. **71**, 225 (2021); Graham, P.W., Irastorza, I.G., Lamoreaux, S.K., Lindner, A., van Bibber, K.A.: Annu. Rev. Nucl. Part. Sci. **65**, 485 (2015)
25. CAST Collaboration, Anastassopoulos, V., et al.: Nat. Phys. **13**, 584 (2017)
26. Meyer, M., Petrushevska, T.: Phys. Rev. Lett. **124**, 231101 (2020) [Erratum ibid. **125**, 119901 (2020)]
27. Payez, A., Evoli, C., Fischer, T., Giannotti, M., Mirizzi, A., Ringwald, A.: JCAP **02**, 006 (2015)
28. Dessert, C., Foster, J.W., Safdi, B.R.: Phys. Rev. Lett. **125**, 261102 (2020)
29. Ivanov, M., et al.: JCAP **02**, 059 (2019)
30. Caputo, A., et al.: Phys. Rev. D **100**, 063515 (2019)
31. Johnson, M.D., et al.: Science **350**, 1242 (2015)
32. Green, M.B., Schwarz, J.H., Witten, E.: Superstring Theory. Cambridge University Press, Cambridge (1987); Hammond, R.: Nuovo Cimento B **109**, 319 (1994); Gen. Relativ. Gravit. **28**, 419 (1986); De Sabbata, V.: Ann. der Phys. **7**, 419 (1991); de Sabbata, V.: Torsion, string tension and quantum gravity. In: "Erice 1992", Proceedings, String Quantum Gravity and Physics at the Planck Energy Scale, p. 528; Murase, Y.: Prog. Theor. Phys. **89**, 1331 (1993)
33. Kubyshin, Yu.A.: J. Math. Phys. **35**, 310 (1994); German, G., Macias, A., Obregon, O.: Class. Quantum Gravity **10**, 1045 (1993); Oh, C.H., Singh, K.: Class. Quantum Gravity **6**, 1053 (1989)
34. Howe, P.S., Opfermann, A., Papadopoulos, G.: Twistor spaces for QKT manifold. Commun. Math. Phys. **197**, 713 (1998); Howe, P.S., Papadopoulos, G.: Phys. Lett. **379B**, 80 (1996); de Sabbata, V., Sivaram, C.: Il Nuovo Cimento A **109**, 377 (1996)
35. Papadopoulos, G., Townsend, P.K.: Nucl. Phys. B **444**, 245 (1995); Hull, C.M., Papadopoulos, G., Townsend, P.K.: Nucl. Phys. B **316**, 291 (1993)
36. De Sabbata, V., Sivaram, C.: Astrophys. Space Sci. **176**, 141 (1991); Goenner, H., Müller-Hoissen, F.: Class. Quantum Gravity **1**, 651 (1984); Minkowski, P.: Phys. Lett. B **173**, 247 (1986); de Ritis, R., Scudellaro, P., Stornaiolo, C.: Phys. Lett. A **126**, 389 (1988); Assad, M.J., Letellier, P.S.: Phys. Lett. A **145**, 74 (1990); Canale, A., de Ritis, R., Tarantino, C.: Phys. Lett. A **100**, 178 (1984); Fennelly, A.J., Smalley, L.L.: Phys. Lett. A **129**, 195 (1988); Buchbinder, I.L., Odintsov, S.D., Shapiro, I.L.: Phys. Lett. B **162**, 92 (1985); Wolf, C.: Gen. Relativ. Gravit. **27**, 1031 (1995); Chatterjee, P., Bhattacharya, B.: Mod. Phys. Lett. A **8**, 2249 (1993)
37. Figueiredo, B.D.B., Damiao Soares, I., Tiomno, J.: Class. Quantum Gravity **9**, 1593 (1992); Garcia de Andrade, L.C.: Mod. Phys. Lett. A **12**, 2005 (1997); Chandia, O.: Phys. Rev. D **55**, 7580 (1997); Letelier, P.S.: Class. Quantum Gravity **12**, 471 (1995); Patricio, S., Letelier, S.: Class. Quantum Gravity **12**, 2221 (1995); Kühne, R.W.: Mod. Phys. Lett. A **12**, 2473 (1997); Ross, D.K.: Int. J. Theor. Phys. **28**, 1333 (1989)
38. Capozziello, S., Stornaiolo, C.: Nuovo Cimento B **113**, 879 (1998)
39. Capozziello, S., Lambiase, G., Stornaiolo, C.: Ann. Phys. **10**, 713 (2001)
40. Cai, Y.-F., Capozziello, S., De Laurentis, M., Saridakis, E.N.: Rep. Prog. Phys. **79**, 106901 (2016)
41. Birrel, N.D., Davies, P.C.W.: Quantum Fields in Curved Space. Cambridge University Press, Cambridge (1982)
42. Hammond, R.T.: Class. Quantum Gravity **13**, 1691 (1996)
43. Itzykson, C., Zuber, J.B.: Quantum Field Theory. McGraw-Hill Inc., New York (1980)
44. Capozziello, S., Iovane, G., Lambiase, G., Stornaiolo, C.: Europhys. Lett. **46**, 710 (1999)

45. Duan, X.-C., Den, X.-B., Zhou, M.-K., Zhang, K., Xu, W.-L., Shao, C.-G., Luo, J., Hu, Z.-K.: Phys. Rev. Lett. **117**, 023001 (2016)
46. Audretsch, J., Lämmerzahl, C.: Appl. Phys. B **54**, 351 (1992)
47. Lämmerzahl, C.: Phys. Lett. A **228**, 223 (1997)
48. Sakurai, J.J.: Advanced Quantum Mechanics. Addison-Wesley, Reading (1967)
49. Berry, M.V.: Proc. R. Soc. Lond. A **392**, 45 (1984)
50. Bakke, K., Nascimento, J.R., Furtado, C.: JHEP **08**, 106 (2008)
51. Beck, Ch.: Phys. Rev. Lett. **111**, 231801 (2013)
52. Hoffmann, C., Lefloch, F., Sanquer, M., Pannetier, B.: Phys. Rev. B **70**, 180503(R) (2004)
53. Golikova, T.E., et al.: Phys. Rev. B **86**, 064416 (2012)
54. Bae, M.-H., Dinsmore, R.C., Sahu, M., Lee, H.-J., Bezryadin, A.: Phys. Rev. B **77**, 144501 (2008)
55. London, F.: Superfluids, vol. 1. Dover Publications, New York (1961)
56. Beck, Ch.: Phys. Dark Universe **7**, 6 (2015)

Perspectives 8

Abstract

The subject of this chapter is the gravitational memory effect, which is a slow-growing, non-oscillatory contribution to the gravitational-wave amplitude. It is generated by GWs that are sourced by the previously emitted waves. A gravitational wave with memory causes a permanent displacement of test masses that persists after the wave has passed. The linear and nonlinear memory effects and the spin memory effect are discussed. Realistic estimates of the detectability of the memory suggest that this effect could be observed by the next generation of GW detectors, such as those being developed for the LISA mission. The gravitational memory induced by neutrino emission in supernova collapse is considered. Quasi-normal Modes (QNMs) for compact stars and black holes are also discussed. They are extremely important in GW astrophysics and play a crucial role in the analysis of the gravitational signal emitted by compact objects. Their eigenfrequencies provide information about the nature and inner structure of the emitting source.

8.1 The Gravitational-Wave Memory Effect

The displacement memory effect is a consequence of the fact that after a Gravitational Wave (GW) has passed through a region of space–time, the latter does not return to its original state, but *keeps a memory* of the GW because the proper distances between objects are permanently changed. This situation persists even after the oscillations have ceased. The effect is visualized in Fig. 8.1 for freely falling masses.

Owing to the memory effect, the standard picture of the GW signals[1] (the waveform from a coalescing compact-object binary) is incomplete. All GW sources

[1] Namely, the oscillatory amplitude which is initially small reaches some maximum, and then decays back to zero at later times.

G. Lambiase and G. Papini, *The Interaction of Spin with Gravity in Particle Physics*, Lecture Notes in Physics 993, https://doi.org/10.1007/978-3-030-84771-5_8

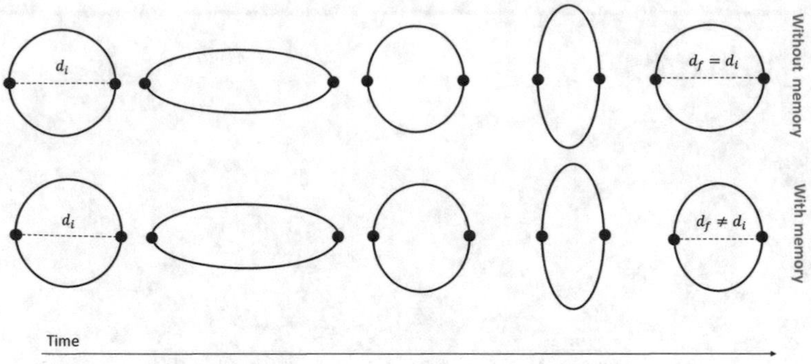

Fig. 8.1 The GW memory effect for a ring of freely falling masses. The first picture shows two particles a distance d_i apart before the passage of a GW. The next three frames show the typical oscillations of the relative positions of the two particles. The final picture shows that their relative positions are permanently changed ($d_f \neq d_i$) after the GW has passed

possess instead some form of *GW memory*, with the property that the late-time and early-time values of the GW polarizations differ from zero [1]

$$\Delta h^{\rm mem}_{+,\times} = \lim_{t\to+\infty} h_{+,\times}(t) - \lim_{t\to-\infty} h_{+,\times}(t), \tag{8.1}$$

where t is time at the observer. Nearby freely falling observers are able to measure the memory effect by means of their geodesic deviation. Since this effect can be produced by any isolated source that radiates energy asymmetrically, it is extremely important to know whether astrophysical sources are able to generate the effect with an amplitude that might be revealed by current or upcoming GW detectors.

Zel'dovich and Polnarev [2] were the first authors to compute the displacement memory arising, in the framework of linearized gravity, from the scattering of stars. Soon after, the displacement memory from gravitational bremsstrahlung was computed in [3–5]. The possibility that neutrino emission from supernovae could generate the displacement memory effect was studied in [4]. Later Christodoulou [6] considered the so-called nonlinear displacement memory, an effect essentially related to the effective energy per solid angle radiated in GWs. The nonlinear displacement memory was also computed for compact binaries by Wiseman and Will [7], and in the post-Newtonian-expanded, multipolar post-Minkowski approximation (PN) by Blanchet and Damour [8], thus confirming that compact binaries do have a nontrivial memory effect. An extension of these calculations to a high order in the PN approximation was carried out by Favata [9]. The displacement memory effect has also been computed by using numerical simulations [10] (see also [11–13]).

Linear, Nonlinear, and Spin Memory Effects

Before giving explicit examples of the linear and nonlinear memory effects, it is necessary to introduce some preliminary concepts used in the formalism developed in these contexts [1]. The GW polarizations are decomposed in a sum over (l, m) modes as

$$h_+ - ih_\times = \sum_{l=2}^{\infty} \sum_{m=-l}^{l} h^{lm} \, {}_{-2}Y^{lm}(\Theta, \Phi), \tag{8.2}$$

where (Θ, Φ) indicate the direction from the source to the observer, and ${}_{-2}Y^{lm}$ are spin-weighted spherical harmonics. The modes h^{lm} in (8.2) and the radiative mass and current multipoles here denoted U^{lm} and V^{lm}, respectively, are related by (in units $G = c = 1$)

$$h^{lm} = \frac{1}{\sqrt{2}r} \left[U^{lm}(T_r) - i V^{lm}(T_r) \right], \tag{8.3}$$

where r is the distance from source to observer, T_r is the retarded time (see also Ref. [9] for details). The moments U^{lm} and V^{lm} can be written in terms of their corresponding symmetric-trace-free (STF) tensors of rank-l, $U^{lm} = A_l \mathcal{U}_L \mathcal{Y}_L^{lm\,*}$ and $V^{lm} = B_l \mathcal{V}_L \mathcal{Y}_L^{lm\,*}$, where A_l and B_l are l-dependent constants, while $\mathcal{Y}_L^{lm}$ are the symmetric-trace-free spherical harmonics, related to the spherical harmonics by $Y^{lm} = \mathcal{Y}_L^{lm} n_L$, with $n_L = n_{i_1} n_{i_2} \cdots n_{i_l}$ a product of l unit radial vectors. The moments U^{lm} and V^{lm} entering (8.2) are related to the source multipole moments $(I_{lm}, J_{lm}, \ldots)$, which are defined in terms of integrals over the stress–energy pseudotensor of the matter and gravitational fields of the source, written using the multipolar post-Minkowski iteration scheme [9] (see [14] for a review). The radiative mass moments U_{lm} are reported here as an example. They are written in terms of the source mass moments I_{lm} and the mass monopole moment $\mathcal{M}$ [1,9]

$$U_{lm} = I_{lm}^{(l)} + 2\mathcal{M} \int_{-\infty}^{T_r} \left[\ln\left(\frac{T_r - \tau}{2\tau_0} \right) + \kappa_l \right] I_{lm}^{(l+2)}(\tau) d\tau + U_{lm}^{(\text{nonlinmem})} + \mathcal{O}(2.5\text{PN}), \tag{8.4}$$

where $I_{lm}^{(l)} = d^l I_{lm}/dt^l$, the integral term is a 1.5PN order tail term, κ_l is a constant depending on l, τ_0 is an arbitrary timescale (it disappears in physical observables), $U_{lm}^{(\text{nonlinmem})}$ is the nonlinear memory term, and $\mathcal{O}(2.5\text{PN})$ accounts for terms entering at 2.5PN as well as higher order terms.

Linear Memory

A typical example of linear memory is given by the waveform from a hyperbolic binary. The polarization (to leading order in multipolar expansion) is

$$h_+ - ih_\times \approx \sum_{m=-2}^{2} \frac{I_{2m}^{(2)}}{\sqrt{2}r} \, {}_{-2}Y^{lm}(\Theta, \Phi). \tag{8.5}$$

Consider a Keplerian binary system in the x-y-plane, where m_1 and m_2 are the masses of the two bodies. Defining by $r(t)$ the relative orbital separation, $M = m_1 + m_2$ the total mass, $\eta \equiv m_1 m_2/M^2$ the reduced mass ratio, and $\varphi(t)$ the orbital phase angle, the mass quadrupole becomes [1]

$$I_{2m} = \frac{16\pi\eta}{5\sqrt{3}} M r^2(t) Y^*_{2m}\left(\frac{\pi}{2}, \varphi(t)\right).$$

Using the Keplerian orbits parameters, hence the semi-latus p, the eccentricity e_0, and the true anomaly $v = \varphi - \omega_p$ (the periastron direction on the x-axis is obtained for $\omega_p = 0$), the orbital motion $r = r(v)$ and the orbital phase angle $\dot{v}$ are given by

$$r = \frac{p}{1 + e_0 \cos v}, \qquad \dot{v} = \dot{\varphi} = \frac{\sqrt{pM}}{r^2}. \tag{8.6}$$

The functions $I^{(2)}_{2m}$ entering the waveforms given in (8.5) are

$$I^{(2)}_{20} = -8\sqrt{\frac{\pi}{15}}\frac{\eta M^2}{p} e_0(e_0 + \cos v), \tag{8.7}$$

$$I^{(2)}_{2\pm2} = -4\sqrt{\frac{2\pi}{5}}\frac{\eta M^2}{p} e^{\mp 2i\varphi(t)}\left[1 - e_0^2 + (1 + e_0\cos v)(1 + 2e_0 e^{\pm iv})\right]. \tag{8.8}$$

The waveforms (8.7) are oscillatory for e_0 in the range $0 \le e_0 < 1$. For $e_0 > 1$ (a hyperbolic orbit) and $\omega_p = 0$, one has at the early times $lim_{t\to-\infty}\varphi = \varphi_- = v_- = -\arccos(-e_0^{-1})$, and at late times $lim_{t\to+\infty}\varphi = \varphi_+ = v_+ = \arccos(-e_0^{-1})$, so that their difference yields

$$\Delta I^{(2)}_{20} = 0,$$

$$\Delta I^{(2)}_{2\pm2} = \pm i16\sqrt{\frac{2\pi}{5}}\frac{\eta M^2}{p}\frac{(e_0^2 - 1)^{3/2}}{e_0^2},$$

which induces the memory in the GW polarization amplitudes. For $e_0 = 1$, that is, for a parabolic orbit, no memory effect occurs owing to the fact that the orbital phase angle returns to its early-time value.

The linear memory for an unbound system can be computed solving the linearized, harmonic gauge Einstein field equations $\Box\bar{h}_{jk} = -16\pi T_{jk}$, where T_{jk} is the stress–energy tensor of N gravitationally unbound particles with masses M_A and constant velocities $\mathbf{v}_A$, $\bar{h}_{jk}$ is the trace-reversed metric perturbation, and $\Box$ is the D'Alambert operator in flat space. The difference between the late- and early-time values of the h_{jk} in the transverse-traceless gauge ($\Delta h^{\rm TT}_{jk} = h^{\rm TT}_{jk}|_{t\to+\infty} - h^{\rm TT}_{jk}|_{t\to-\infty}$) is [15,16]

$$\Delta h^{\rm TT}_{jk} = \Delta \sum_{A=1}^{N} \frac{4M_A}{R\sqrt{1 - v_A^2}}\left[\frac{v_A^j v_A^k}{1 - \mathbf{v}_A \cdot \mathbf{N}}\right]^{\rm TT}. \tag{8.9}$$

In (8.9), the masses and velocities could refer to the pieces of a disrupted binary, to a gamma-ray-burst jet [17], to the radiated neutrinos [4,18], and to pieces of ejected material in a supernova explosion [19–21].

Nonlinear Memory

The nonlinear memory (or Christodoulou memory) arises from the energy flux of the radiated GWs that contribute to the radiative mass multipole moments U_{lm} [6,8, 22]. Consider the Einstein field equations. In the harmonic gauge, one has $\Box\bar{h}^{\alpha\beta} = -16\pi\tau^{\alpha\beta}$, where $\tau^{\alpha\beta}$ includes the stress–energy tensor of matter $T^{\alpha\beta}$, the Landau–Lifshitz pseudotensor $t_{\rm LL}^{\alpha\beta}$, and other terms quadratic in $\bar{h}^{\alpha\beta}$ [23]. The pseudotensor $t_{\rm LL}^{\alpha\beta}$ contains a term that is proportional to the stress–energy tensor associated to GWs, namely, $T_{jk}^{GW} = \frac{\mathcal{F}}{r^2} n_j n_k$, where n_j denotes a unit radial vector and $\mathcal{F} \equiv \frac{dE^{GW}}{dt d\Omega}$ is the energy flux of the GWs. This term induces a correction $\delta h_{jk}^{\rm TT}$ to the GW field given by [7]

$$\delta h_{jk}^{\rm TT} = \frac{4}{r}\int_{-\infty}^{T_r} dt' \left[\int \frac{dE_{GW}}{dt' d\Omega'} \frac{n'_j n'_k}{(1 - \mathbf{n}' \cdot \mathbf{N})} d\Omega'\right]^{\rm TT}. \tag{8.10}$$

The time integral is responsible for the memory effects in what the GW field, for any value of T_r, depends on the past history of the source. Interestingly, the nonlinear memory (8.10) can be cast in terms of the linear memory (8.9) in the case in which the unbound objects in the system individually radiate gravitons with energies $E_A = M_A/(1 - v_A^2)^{1/2}$ and velocities $v_A^j = c\, n_A'^j$ [16].

The decomposition (8.2) allows to write the nonlinear memory correction to the radiative mass multipole moments U_{lm} (8.4) (the radiative current moments V_{lm} instead do not have a nonlinear memory contribution) in the form [8,9]

$$U_{lm}^{\rm (nonlinmem)} = 32\pi \sqrt{\frac{(l-2)!}{2(l+2)!}} \int_{-\infty}^{T_R} dt \int d\Omega \frac{dE_{GW}}{dt d\Omega}(\Omega) Y_{lm}^*(\Omega), \tag{8.11}$$

where

$$\frac{dE_{GW}}{dt d\Omega} = \frac{r^2}{16\pi}\langle \dot{h}_+^2 + \dot{h}_\times^2 \rangle = \frac{r^2}{16\pi} \sum_{l',l'',m',m''} \langle \dot{h}_{l'm'} \dot{h}_{l''m''}^* \rangle {}_{-2}Y^{l'm'}(\Omega) {}_{-2}Y^{l''m''*}(\Omega). \tag{8.12}$$

These results show that the memory is calculated iteratively, in the sense that h_{lm} modes that contain the memory are negligible in the computation of the energy flux.

Spin Memory Effects

Another type of gravitational-wave memory effect is spin memory. This is generated by asymmetric changes in the angular momentum (per unit solid angle) to null infinity in massless fields [24,25]. The spin memory is also produced by the variation in superspin charges, which is the magnetic-parity part of the charges conjugate to the super-rotation vector fields.[2]

In the Bondi framework, one uses the set of coordinates (u, r, θ^A), where u is a retarded time, r is an affine parameter along outgoing null rays, and θ^A, with $A = 1, 2$, are arbitrary coordinates defined on the 2-sphere. The components of the metric can be written in the form [25]

$$ds^2 = -\left(1 - \frac{2m}{r}\right) du^2 + 2du dr + D^B C_{AB} d\theta^A du + r^2 \left(h_{AB} + \frac{1}{r} C_{AB}\right) d\theta^A d\theta^B + \mathcal{O}\left(\frac{1}{r}\right). \tag{8.13}$$

Here h_{AB} is the metric on the 2-sphere, D_A is the covariant derivative defined in terms of h_{AB}, $C_{AB}(u, \theta^C)$ is the shear tensor, and $m(u, \theta^A)$ is the Bondi mass aspect. The tensor C_{AB} is symmetric and trace-free, and contains information about the usual gravitational-wave strain h_{ij}^{TT}, and can be written as a sum of two terms

$$\begin{aligned} C_{AB} &= C^{\mathrm{E}}_{AB} + C^{\mathrm{B}}_{AB} \\ C^{\mathrm{E}}_{AB} &\equiv \frac{1}{2}(2D_A D_B - h_{AB} D^2)\Phi \\ C^{\mathrm{B}}_{AB} &\equiv \epsilon_{C(A} D_{B)} D^C \Psi \,, \end{aligned} \tag{8.14}$$

where ϵ_{AB} is the antisymmetric tensor. The functions Φ and Ψ are smooth functions of the coordinates (u, θ^A). The tensor C^{E}_{AB} is essentially related to the gravitational memory effect discussed in the previous sections, while C^{B}_{AB} is related to the spin memory effect.

The relevant quantity for computing the spin memory is the observable quantity defined by [25,30]

$$\Delta\Sigma \equiv \int_{-\infty}^{+\infty} du \Psi(u, \theta^A). \tag{8.15}$$

The quantity $\Delta\Sigma$ in (8.15) can be derived from the variation of the flux of angular momentum per unit solid angle in GWs and matter, as well as from the variation

[2]The displacement memory is also related to the symmetry group of asymptotically flat space–times, the Bondi–Metzner–Sachs (BMS) group [26–28]. The BMS group has a similar structure to the Poincaré group: it is the semidirect product of an Abelian group with a group isomorphic to the proper, orthochronous Lorentz group. It is characterized by having an infinite-dimensional group of supertranslations (which include the four-dimensional group of ordinary translations), rather than the usual four-dimensional group of translations. The displacement memory is related to this supertranslation which connects a frame at late times to a frame at early times [29–31] (a similar formulation can be expressed in terms of the transformation needed to transform a preferred Poincaré group at early times to a different preferred Poincaré group at late times [32]). Notice, finally, that the changes in charges conjugate to the supertranslations (the supermomentum charges) generate the displacement memory, as well.

of the curl of a quantity related to the generalization of the spin of the system (the superspin charges).

Expression (8.15) simplifies considerably, in the post-Minkoskian approximation, for non-spinning, quasicircular binaries. Consider the case of a binary with orbital angular momentum in the z-direction. The relevant multipoles are $U_{2,2}$ and $\dot{U}_{2,-2}$, given by

$$U_{2,2} = -8\sqrt{\frac{2\pi}{5}}\,\eta M x\, e^{ix^{-5/2}/16\eta} + \mathcal{O}\left(\frac{1}{c^2}\right)$$

$$\dot{U}_{2,2} = 16i\sqrt{\frac{2\pi}{5}}\,\eta M x^{5/2}\, e^{ix^{-5/2}/16\eta} + \mathcal{O}\left(\frac{1}{c^2}\right)$$

and their complex conjugates, while the relevant moment to obtain the spin memory is the $\{l = 3, m = 0\}$ mode. All other modes are higher post-Minkoskian quantities. The quantity (8.15) assumes the form

$$\Delta\Sigma = \frac{Y_{3,0}}{80\sqrt{7\pi}}\int_{-\infty}^{u_f} du\,\Im(\bar{U}_{2,2}\dot{U}_{2,2}) + \mathcal{O}\left(\frac{1}{c^2}\right) \tag{8.16}$$
$$= \frac{1}{10}\sqrt{\frac{\pi}{7}}\eta M^2(x_f^{-1/2} - x_{-\infty}^{-1/2})Y_{3,0} + O(c^{-2}).$$

Here $\Im(...)$ denotes the imaginary part. The integration of C^{B}_{AB} over u gives $\int du C^{\mathrm{B}}_{AB} = \epsilon_{C(A}D_{B)}D^C\Delta\Sigma$, which can be explicitly computed to give

$$\epsilon_{C(A}D_{B)}D^C\Delta\Sigma = \sqrt{\frac{3\pi}{35}}\eta M^2(x_f^{-1/2} - x_{-\infty}^{-1/2})T^{(b),3,0}_{AB} + O\left(\frac{1}{c^2}\right), \tag{8.17}$$

where $T^{(b),3,0}_{AB}$ is related to $Y_{3,0}$ by the general relation

$$T^{(b),l,m}_{AB} = \sqrt{\frac{2(l-2)!}{(l+2)!}}\,\epsilon_{C(A}D_{B)}D^C Y_{l,m}.$$

The contribution of the spin memory to the u integral of the strain is given by the cross component

$$\int_{-\infty}^{u_f} du h^{\mathrm{smm}}_{\times} = \frac{1}{r}\sqrt{\frac{3\pi}{70}}\eta M^2(x_f^{-1/2} - x_{-\infty}^{-1/2})_{-2}Y_{3,0} + \mathcal{O}\left(\frac{1}{c^2}\right)$$
$$= \frac{3}{8r}M^2\eta(x_f^{-1/2} - x_{-\infty}^{-1/2})\sin^2\theta\cos\theta + \mathcal{O}\left(\frac{1}{c^2}\right). \tag{8.18}$$

By differentiating (8.18), one gets

$$h^{\mathrm{smm}}_{\times} = -\frac{12M\eta^2}{5r}x^{7/2}\sin^2\theta\cos\theta + \mathcal{O}\left(\frac{1}{c^2}\right). \tag{8.19}$$

This expression does coincide with the one derived in [33]. In terms of the radiative mass quadrupole moments $U_{2,2}$, hence in a more general form, Eq. (8.19) reads

$$h_{\times}^{\text{smm}} = \frac{3}{64\pi r}\Im(\bar{U}_{2,2}\dot{U}_{2,2})\sin^2\theta\cos\theta\,. \tag{8.20}$$

This part of $h_\times$ is intimately connected to the spin memory, and therefore $h_{\times}^{\text{smm}}$ is referred to in the literature as the spin memory mode (SMM).

Gravitational Memory Induced by Neutrino Emission

An interesting aspect that deserves to be mentioned is the role that GWs and neutrinos (including electromagnetism) will have in the wave astronomy, which constitutes the new frontiers of multi-messenger astronomy. In fact, it could provide important answers about the core collapse of massive stars and their implosion into black holes, as well as the processes of inspiral and merger of a binary systems. Although neutrino and GW physics developed separately, their interplay is now extremely important. One of the consequences is the gravitational memory induced by the (anisotropic) emission. The anisotropic emission is controlled by the anisotropy parameter $\alpha(t)$,

$$\alpha(t) = \frac{1}{L_\nu(t)}\int_{4\pi} d\Omega'\,\Psi(\vartheta',\varphi')\,\frac{dL_\nu(\Omega',t)}{d\Omega'}\,, \tag{8.21}$$

where $\Psi(\vartheta',\varphi')$ accounts for the location of the observer with respect to the source [34,35]. The anisotropy parameter is important for determining the amplitude of the gravitational-wave strain $h(t)$, which is related to it by the relation (see [36])

$$h(t) = \frac{2G}{rc^4}\int_{-\infty}^{t-r/c} dt' L_\nu(t')\alpha(t')\,, \tag{8.22}$$

where $L_\nu(t')$ is the neutrino luminosity. For a finite duration of the neutrino burst ($\Delta t \sim 10$ s), one expects a transition of metric perturbation $h(t)$ from an asymptotic value $h(t\to-\infty)=0$ to a different asymptotic value $h(t\to+\infty)=\Delta h$. The gravitational memory accumulates from the arrival of the first neutrino to Earth until the neutrino burst has passed completely. An estimate of the strain h is given by

$$\begin{aligned}|h(t)| &\le \frac{2G}{rc^4}\int_{-\infty}^{\infty} L_\nu(t)|\alpha(t)|dt \le \frac{2G}{rc^4}|\alpha|_{max}E_{tot}\,, \\ &\le 6.41\times10^{-20}\frac{|\alpha|_{max}}{0.04}\,\frac{E_{tot}}{3\;10^{53}\text{ ergs}}\,\frac{10\text{kpc}}{r}\,,\end{aligned} \tag{8.23}$$

where $E_{tot} = \int_{-\infty}^{\infty} L_\nu(t)dt \simeq 3\times10^{53}$ ergs is the total energy emitted as neutrinos.

In a GW detector, the neutrino-induced memory from a galactic supernova would appear as a signal with typical frequency of (0.1–10) Hz and (dimensionless) strain of

$h \sim (10^{-22}–10^{-20})$. This signal is below the sensitivity of LIGO and its immediate successors, but could be observed with the third generation of GW detectors, in particular, those able to explore the Deci-Hz frontier [37–42] (the region centred at $f \sim 0.1$Hz), such as DECIGO (DECi-hertz Gravitational-wave Observatory) [37, 38] and BBO (Big Bang Observer) [38], which will reach a sensitivity of 10^{-24} in strain. These detectors will allow, indeed, to explore the supernova neutrino memory, providing another important confirmation of GR.

8.2 Quantum Wave Equations and Quasi-normal Modes

Isolated black holes (BHs) (in equilibrium) are described by few parameters: their mass M, angular momentum L, and charge Q. However, BHs are not isolated. In fact, BHs in centres of galaxies or BHs with intermediate mass are surrounded by complex distributions of matter, such as galactic nuclei, accretion discs, strong magnetic fields, and so on, with which they interact actively. Moreover, even removing the macroscopic objects and fields around a BH, the latter may interact with the vacuum around it, create pairs of particles, and evaporate via the Hawking radiation process. Therefore, in a real situation, BHs cannot be described by only their basic parameters $\{M, L, Q\}$, but they are always in the perturbed state. A perturbed BH metric can be described in terms of the unperturbed background metric $g^0_{\mu\nu}$, namely,

$$g_{\mu\nu} = g^0_{\mu\nu} + \delta g_{\mu\nu}, \tag{8.24}$$

where, in linear approximation, $\delta g_{\mu\nu}$ is the metric perturbations (supposed to be much less than the background metric, $\delta g_{\mu\nu} \ll g^0_{\mu\nu}$). The background metric $g^0_{\mu\nu}$ can be, for example, the Schwarzschild or Kerr metrics.

As a consequence of the perturbation, a BH emits GWs with a time evolution characterized by three stages: (1) a (relatively) short period of initial outburst of radiation; (2) a long period of damping proper oscillations (this period is dominated by the QNMs); and (3) at very large time the QNMs are suppressed by power law or exponential late-time tails. QNMs (here "quasi" refers to the fact that the system is open and loses energy through gravitational radiation) play a crucial role[3] because the new generation of gravitational antennas could be able to detect such a kind of gravitational signal from BHs. In most cases, QNMs refer to the frequencies of QNMs and not to the corresponding amplitudes.

[3]It is worth to mention that QNMs are also important in the context of the duality between supergravity in anti-de Sitter space–time (AdS) and conformal field theory (CFT), the AdS/CFT correspondence [43,44].

Table 8.1 Wave equations for the scalar, spinor, Proca, and Maxwell fields

Field	Spin	Wave equation
Scalar	$s = 0$	$\left[\frac{1}{\sqrt{-g}}\partial_\mu\left(\sqrt{-g}g^{\mu\nu}\partial_\nu\right) - m^2\right]\Psi = 0$
Spinor	$s = 1/2$	$\left[\gamma^a e_a^{\ \mu}(\partial_\mu + \Gamma_\mu) + m\right]\Psi = 0$
Proca	$s = 1$	$\frac{1}{\sqrt{-g}}\partial_\mu\left(\sqrt{-g}g^{\mu\rho}g^{\nu\sigma}F_{\rho\sigma}\right) - m^2 A^\mu = 0$
Maxwell	$s = 1$	$\frac{1}{\sqrt{-g}}\partial_\mu\left(\sqrt{-g}g^{\mu\rho}g^{\nu\sigma}F_{\rho\sigma}\right) = 0$

Master Wave Equations and Properties of QNMs

Perturbations of a BH space–time are treated either by adding fields to the black hole space–time or by perturbing the background BH metric. In the linear approximation (fields do not induce back-reaction effects on the background), the first type of perturbation is reduced to the propagation of fields in the background of a BH. The covariant form of the equation of motion of fields depends on the spin s. Table 8.1 refers to the wave equations discussed in the previous chapters.

Consider the covariant Klein–Gordon equation for a scalar field Φ of mass μ in the background of the Schwarzschild geometry $g^0_{\mu\nu}$: $g_{00} = -g_{rr}^{-1} = 1 - 2M/r$, $g_{\theta\theta} = g_{\phi\phi}\sin^{-2}\theta = r^2$,

$$\frac{1}{\sqrt{-g}}\partial_\nu\left(g^{\mu\nu}\sqrt{-g}\partial_\mu\Psi\right) - \mu^2\Psi = 0 \quad (s = 0). \tag{8.25}$$

The purpose of the analysis of the BH perturbation equations is to reduce Eq. (8.25) to the two-dimensional wave-like form. To this aim, one decouples the angular variables, and writes the wave function in the form

$$\Psi(t, r, \theta, \phi) = \frac{R(r)}{r} Y_\ell(\theta, \phi)\, e^{-i\omega t}.$$

The spherical harmonics fulfill the equation

$$\Delta_{\theta,\phi} Y_\ell(\theta, \phi) = -\ell(\ell + 1) Y_\ell(\theta, \phi),$$

while for the radial part, introducing the new variable $dz = \frac{dr}{1-2M/r}$ (or $z = r + 2M\log\left(\frac{r}{2M} - 1\right)$) one gets

$$-\frac{d^2R}{dz^2} + V(r, \omega)R = \omega^2 R, \tag{8.26}$$

where

$$V(r) = \left(1 - \frac{2M}{r}\right)\left[\frac{\ell(\ell + 1)}{r^2} + \frac{2M(1 - s^2)}{r^3}\right]. \tag{8.27}$$

For $s = 0$, (8.27) corresponds to the scalar field, $s = 1$ to the Maxwell field, and $s = 2$ to the gravitational perturbations of the axial type [45]. It must be pointed out that the separation of variables is quite involved, since, in general, the variables in perturbation equations cannot be decoupled for perturbations of an arbitrary metric. In the latter case, in fact, the metric must possess sufficient symmetry related to the existence of the Killing vectors, Killing tensors, and Killing–Yano tensors [46,47] (see also [48,49]).

QNMs are solutions of the wave equation (8.26), satisfying specific boundary conditions at the BH horizon and far from the BH. These are given by[4] (required by causality)

$$\Psi \sim e^{-i\omega z} \text{ (pure ingoing wave)} \quad z \to -\infty. \tag{8.29}$$

$$\Psi \sim e^{+i\omega z} \text{ (pure outgoing wave)} \quad z \to +\infty. \tag{8.30}$$

Since $\Psi \sim e^{-i\omega t}$, the QNM frequencies can be written in the form

$$\omega = \omega_R - i\omega_I\,, \tag{8.31}$$

where ω_R is the real oscillation frequency of the mode and ω_I is proportional to its damping rate ($\omega_I > 0$ means that Ψ is damped, $\omega_I < 0$ means an instability). The quasi-normal frequencies do not depend on a way by which the black hole or a field around it was perturbed. Thus, QNMs are completely determined by BH parameters.

The backscattering effect of the effective potential (8.27) does not allow the calculation of the eigenvalues ω in closed form, but numerically [51–53]. Referring to the fundamental $l = 2$ mode (the lowest dynamical multipole in GR) of gravitational perturbations, one gets

$$M\omega \equiv M(\omega_R + i\omega_I) \approx 0.373672 - i0.0889623\,, \tag{8.32}$$

which are complex (called quasi-normal frequencies). The imaginary components describe the decay in time of fluctuations, with a characteristic timescale $\tau \equiv 1/|\omega_I|$. Since the system is open, waves can travel to infinity or down the horizon and therefore it is physically sensible that any fluctuation damps down (the corresponding modes are the QNMs, and in general do not form a complete set). The QNM spectrum is determined by appropriate boundary conditions, which play a crucial

[4]In string theory, the boundary condition at infinity is that the wave function Ψ must vanish at infinity, while for higher spin fields some gauge-invariant combination (GIC) of field components, dictated by the AdS/CFT correspondence [50], must vanish. Thus, in string theory, the boundary condition at infinity is that of Dirichlet

$$\begin{aligned} \Psi \to 0, &\quad z \to \infty \quad (s = 0), \\ GIC \to 0, &\quad z \to \infty \quad (higher\ s). \end{aligned} \tag{8.28}$$

.

role. Assume that the reflective surface is placed at $r_0 = 2M(1+\epsilon) \gtrsim 2M$, and impose the Dirichlet or Neumann boundary conditions at $r = r_0$. Here $r = r_0(1+\epsilon)$ refers to ultra-compact objects. In the $\epsilon \to 0$ limit, the QNMs are given by [54]

$$M\omega_R \simeq \frac{M}{2|z_0|}(p\pi - \delta) \sim |\log\epsilon|^{-1}\,, \tag{8.33}$$

$$M\omega_I \simeq -\beta_{ls}\frac{M}{|z_0|}(2M\omega_R)^{2l+2} \sim -|\log\epsilon|^{-(2l+3)}\,. \tag{8.34}$$

Here $z_0 \equiv z(r_0) \sim 2M\log\epsilon$, p is an odd (even) integer for Dirichlet (Neumann) boundary conditions, δ is the phase of the wave reflected at $r = r_0$, and $\beta_{ls} = \left[\frac{(l-s)!(l+s)!}{(2l)!(2l+1)!!}\right]^2$ [55,56]. Such a result follows from the fact that low-frequency waves are almost trapped by the potential (8.27), so that their frequency scales as $\omega_R \sim 1/z_0$, that is, they as the the size of the cavity (in tortoise coordinates), while the (small) imaginary part arises because waves tunnel through the potential and reach infinity, with a tunnelling probability $|\mathcal{A}|^2 \sim (M\omega_R)^{2l+2} \ll 1$ (this formula holds in the regime of small frequency and scales [55]) [57–60]. The wave trapped in the region is reflected N times. Indicating with t the duration of the reflection, and with z_0 the dimension of the region, one gets $N = t/z_0$ times. The amplitude of the wave reduces to $A(t) = A_0\left(1 - |\mathcal{A}|^2\right)^N \sim A_0\left(1 - t|\mathcal{A}|^2/z_0\right)$. On the other hand, the amplitude is given by $A(t) \sim A_0 e^{-|\omega_I|t} \sim A_0(1 - |\omega_I|t)$, and in this limit, one obtains

$$\omega_R \sim \frac{1}{z_0}\,, \qquad \omega_I \sim \frac{|\mathcal{A}|^2}{z_0} \sim \omega_R^{2l+3}\,, \quad \forall l\,. \tag{8.35}$$

This scaling agrees for any type of perturbation. Here one is assuming that the surface is perfectly reflecting, but in some models it is shown that the low-frequency waves are reflected, whereas higher frequency waves probe the internal structure of the objects [61,62].

For a BH, the excitation of the space–time modes occurs at the photosphere [63], so that the vibrations travel to infinity (to observers) or down the event horizon (the pulse crosses the photosphere, and excites its modes). The ringdown signal is described by the lowest QNMs, and a fraction of this signal reaches the observers, while the other fraction of the signal travels downwards and into the horizon. For compact objects, on the contrary, the pulse travelling inwards is semi-trapped between the object and the light ring. It happens that, interacting with the light ring, a fraction of the pulse travels outside to observers, generating a series of *echoes*. Such echoes are characterized by ever-decreasing amplitudes, with an *echo delay time* (the time of repeated reflections) given by [57,58] (see Eqs. (8.33) and (8.34))

$$\tau_{\rm echo} \sim 4M|\log\epsilon|\,. \tag{8.36}$$

Precise measurements of the ringdown frequencies and damping times are important because allow to test, in binary systems, whether or not the object is a BH [64, 65]. Such analysis can be extended to a large class of objects (for example, boson stars, gravastars, wormholes, or other quantum-corrected objects). In the case of rotating sources, the Kerr horizon is replaced by a reflective surface at $r_0 = r_+(1+\epsilon)$, where $r_\pm = M(1 \pm \sqrt{1-\chi^2})$ are the locations of the horizons and $\chi = J/M^2$ is the dimensionless spin. In Eq. (8.33), ω_R is replaced by $\omega_R \to \omega_R - m\Omega$, where Ω is the angular velocity of the object when $\epsilon \to 0$, whereas Eq. (8.34) assumes the form

$$M\omega_I \simeq -\frac{\beta_{ls}}{|z_0|}\left(\frac{2M^2 r_+}{r_+ - r_-}\right)\left[\omega_R(r_+ - r_-)\right]^{2l+1}(\omega_R - m\Omega)\,, \tag{8.37}$$

with $z_0 \sim M[1 + (1-\chi^2)^{-1/2}]\log\epsilon$. The angular momentum can bring substantial changes, for example, the space–time is unstable for $\omega_R(\omega_R - m\Omega) < 0$ (i.e. in the superradiant regime), on a timescale $\tau_{\text{inst}} \equiv 1/\omega_I$. This phenomenon is called *ergoregion instability* [56,66,67]. Moreover, the echo delay time (8.36) is given by $\tau_{\text{echo}} \sim 2M[1 + (1-\chi^2)^{-1/2}]\log\epsilon$.

Finally, the nature of inspiralling objects, and how they respond to the external gravitational field of their companion requires some discussion. This analysis is performed in terms of the tidal Love numbers k [68]. In the case of BH, described by General Relativity, the tidal Love number is exactly zero [69–71], whereas, for exotic compact objects (ECOs), they are small but finite [72]. In particular, the tidal Love numbers of ClePhOs vanish logarithmically in the BH limit. Thus, a measurement of the tidal Love number k provides an estimation of the distance of the ECO surface from its Schwarzschild radius, $\epsilon \sim e^{-1/k}$. The tidal Love number k of an ECO of the order $\sim \mathcal{O}(10^{-3})$ allows to probe Planck distances away from the gravitational radius r_g [72], which could be, in principle, within the accuracy of future GW detectors [73].

Excellent reviews on QNMs and related topics here discussed can be found in [52,53,74–78].

8.3 Conclusions

Owing to the huge number of astrophysical sources that may generate displacement memory, the observation of the latter is now a developing research line in the study of GWs, thanks also to the improvement in the sensitivity achieved by GW detectors. It is expected that future ground-based GW detectors will be able to observe the displacement memory. Space-based interferometers, such as the LISA (Laser Interferometer Space Antenna) mission [79], could have the sensitivity needed to detect the memory signal from individual supermassive-binary-black-hole mergers [9]. The LISA project involves three spacecrafts centred around a freely falling test mass, forming an equilateral triangle with arms of length $\sim 5 \times 10^6$km. It will be trailing Earth by $\sim 1AU$ (see Fig. 8.2). In comparison with LIGO, which is sensitive to a frequency range of about 10–1000Hz, the space-based observatory LISA will have a lower and broader frequency range, about 10^4–1 Hz. This means that LISA mission will have the potential to detect events from the merging of supermassive binaries with masses $(10^5–10^7)M_\odot$.

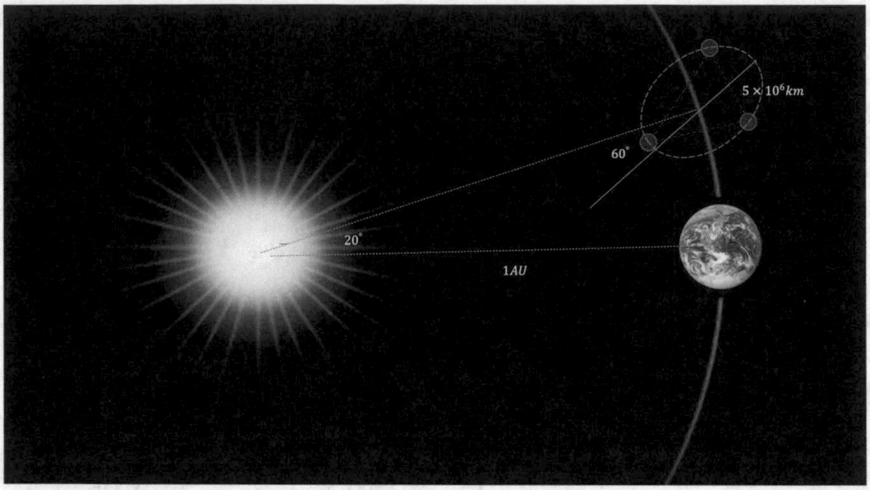

Fig. 8.2 The schematic LISA mission

QNMs certainly represent one of the most important approaches to the understanding of the dynamics of astrophysical objects, such as BHs (including black branes) and exotic compact objects. QNMs are single frequency modes that dominate the time evolution of perturbations of systems that undergo internal dissipation or emission of energy. Owing to these damping effects, the frequency of a QNM must be complex (the characteristic damping time is related to the imaginary part of the QNM frequency). In astrophysical frameworks, GW measurements of QNMs provide information on BH masses and spins (and more generally, both the classical and quantum properties), with unprecedented accuracy, will allow to test the no-hair theorem of general relativity an [52,53,74–78]. Future connections among QNM research, fundamental physics, and astrophysics will certainly constitute the research frontier in the next few years.

The selection of topics covered by this chapter should not be understood as an indication of an exclusive direction in which research has progressed. In fact, several theoretical advances of note should be mentioned. Though they are the subject of intense research particularly in string theory, the possibility that they yield important results should not be underestimated. At the risk of doing an injustice to these developments, brief mention must be given of holography, "double-copy" and Kawai–Lewellen–Tye relations.

Research in holography predicts that one of the space dimensions could be an illusion and that all particles and fields that are part of reality in actuality wonder about in two dimensions. Gravity would not be a force present in two dimensions, but would emerge with the appearance of the third dimension. The advantage of this view is that problems that are mathematically intractable in three dimensions may be soluble in lower dimensions [80–87].

Prime examples of usefulness of string theory in the study of field theoretical amplitudes are the Kawai–Lewellen–Tye (KLT) relations discovered in 1985 [88].

They provide a way of writing scattering amplitudes of closed strings in terms of quadratic combinations of open string amplitudes. In the field theory limit where closed strings reduce to gravitons (particle excitations of general relativity) and open strings to gluons (excitations of Yang–Mills theory), KLT relations give connections between gravitons and gluon scattering amplitudes. Such relations not only hint a relation between the two theories, but also provide simplifications for practical calculations [89–91].

The double copy approach asks the question whether gravity can be viewed in some sense as the product of two gauge theories. Advanced in the past in connection with electromagnetism [92,93] and gravity [94,95], this notion was realized concretely at the level of tree-level scattering amplitudes by means of the KLT relations of string theory: closed strings tree amplitudes can be written as sums over products of open string amplitudes [96].

The role of the spin is important to understand gravity in the framework of quantum mechanics, and, in particular, whether gravity is a quantum entity. Despite several proposals, testing the quantum nature of gravity in laboratory experiments still seems in the future. The authors of some recent papers [97,98] introduce an interesting idea based on the principle that two objects cannot be entangled without a quantum mediator, showing that, despite the weakness of gravity, *the phase evolution induced by the gravitational interaction of two micron size test masses in adjacent matter–wave interferometers can detectably entangle them even when they are placed far apart. Witnessing* this entanglement satisfies a condition that allows the testing of gravity as a quantum coherent mediator, by means of spin correlation measurements.

Prominent attempts to merge the world of quantum particles with that of gravity like Loop Quantum Gravity and String Theory are well-established research areas that have been reviewed in countless publications [99,100].

References

1. Favata, M.: Class. Quantum Gravity **27**, 084036 (2010)
2. Zel'Dovich, Y.B., Polnarev, A.G.: Astron. Zh. **51**, 30 (1974); Zel'Dovich, Y.B., Polnarev, A.G.: Sov. Astron. **18**, 17 (1974)
3. Turner, M.S.: Astrophys. J. **216**, 610 (1977)
4. Turner, M.S.: Nature (London) **274**, 565 (1978)
5. Kovacs, S.J., Jr., Thorne, K.S.: Astrophys. J. **224**, 62 (1978)
6. Christodoulou, D.: Phys. Rev. Lett. **67**, 1486 (1991)
7. Wiseman, A.G., Will, C.M.: Phys. Rev. D **44**, R2945 (1991)
8. Blanchet, L., Damour, T.: Phys. Rev. D **46**, 4304 (1992)
9. Favata, M.: Phys. Rev. D **80**, 024002 (2009)
10. Pollney, D., Reisswig, C.: Astrophys. J. Lett. **732**, L13 (2011)
11. Cao, Z., Han, W.: Class. Quantum Gravity **33**, 155011 (2016)
12. Mitman, K., Moxon, J., Scheel, M.A., Teukolsky, S.A., Boyle, M., Deppe, N., Kidder, L.E., Throwe, W.: Phys. Rev. D **102**, 104007 (2020)
13. Johnson, A.D., Kapadia, S.J., Osborne, A., Hixon, A., Kennefick, D.: Phys. Rev. D **99**, 044045 (2019)
14. Blanchet, L.: Living Rev. Relativ. **9**, 4 (2006)
15. Braginsky, V.B., Thorne, K.S.: Nature (London) **327**, 123 (1987)

16. Thorne, K.S.: Phys. Rev. D **45**, 520 (1992)
17. Sago, N., Ioka, K., Nakamura, T., Yamazaki, R.: Phys. Rev. D **70**, 104012 (2004)
18. Epstein, R.: Astrophys. J. **223**, 1037 (1978)
19. Burrows, A., Hayes, J.: Phys. Rev. Lett. **76**, 352 (1996)
20. Ott, C.D.: Class. Quantum Gravity **26**, 063001 (2009)
21. Murphy, J.W., Ott, C.D., Burrows, A.: Astrophys. J. **707**, 1173 (2009)
22. Payne, P.N.: Phys. Rev. D **28**, 1894 (1983)
23. Thorne, K.S.: Rev. Mod. Phys. **52**, 299 (1980)
24. Flanagan, E.E., Nichols, D.A., Stein, L.C., Vines, J.: Phys. Rev. D **93**, 104007 (2016); Flanagan, E.E., Nichols, D.A.: Phys. Rev. D **92**, 084057 (2015)
25. Nichols, D.A.: Phys. Rev. D **95**, 084048 (2017)
26. Bondi, H., van der Burg, M.G.J., Metzner, A.W.K.: Proc. R. Soc. A **269**, 21 (1962)
27. Sachs, R.K.: Proc. R. Soc. A **270**, 103 (1962)
28. Sachs, R.: Phys. Rev. **128**, 2851 (1962)
29. Strominger, A., Zhiboedov, A.: J. High Energy Phys. **01**, 086 (2016)
30. Flanagan, E.E., Nichols, D.A.: Phys. Rev. D **95**, 044002 (2017)
31. Bhattacharjee, S., Kumar, S.: Phys. Rev. D **102**, 044041 (2020); Bhattacharjee, S.T., Kumara, S., Bhattacharyya, A.: JHEP **03**, 134 (2021)
32. Ashtekar, A.: Geometry and physics of null infinity. In: Bieri, L., Yau, S.T. (eds.) "Surveys in Differential Geometry", a Jubilee Volume on General Relativity and Mathematics Celebrating 100 Years of General Relativity. arXiv:1409.1800
33. Arun, K.G., Blanchet, L., Iyer, B.R., Qusailah, M.S.S.: Class. Quantum Gravity **21**, 3771 (2004) [Erratum Class. Quantum Gravity **22**, 3115(E) (2005)]
34. Mueller, E., Janka, H.T.: AAP **317**, 140 (1997); Muller, E., Janka, H.T., Wongwathanarat, A.: Astron. Astrophys. **537**, A63 (2012); Walk, L., Tamborra, I., Janka, H.-T., Summa, A., Kresse, D.: Phys. Rev. D **101**, 123013 (2020); Tamborra, I., Hanke, F., Müller, B., Janka, H.-T., Raffelt, G.: Phys. Rev. Lett. **111**, 121104 (2013); Tamborra, I., Raffelt, G., Hanke, F., Janka, H.-T., Mueller, B.: Phys. Rev. D **90**, 045032 (2014); Walk, L., Tamborra, I., Janka, H.-T., Summa, A.: Phys. Rev. D **98**, 123001 (2018); Walk, L., Tamborra, I., Janka, H.-T., Summa, A.: Phys. Rev. D **100**, 063018 (2019)
35. Kotake, K., Ohnishi, N., Yamada, S.: Astrophys. J. **655**, 406 (2007); Kotake, K., Iwakami, W., Ohnishi, N., Yamada, S.: Astrophys. J. **704**, 95 (2009)
36. Mukhopadhyay, M., Cardona, C., Lunardini, C.: arXiv:2105.05862v1 [astro-ph.HE]
37. Seto, N., Kawamura, S., Nakamura, T.: Phys. Rev. Lett. **87**, 221103 (2001)
38. Yagi, K., Seto, N.: Phys. Rev. D **83**, 044011 (2011)
39. Luo, J., Chen, L.-S., Duan, H.-Z., Gong, Y.-G., Hu, S., Ji, J., et al.: Class. Quantum Gravity **33**, 035010 (2016)
40. Graham, P.W., Hogan, J.M., Kasevich, M.A., Rajendran, S.: Phys. Rev. D **94**, 104022 (2016)
41. Shuichi Sato, S.K., Masaki Ando, E.: J. Phys.: Conf. Ser. **840**, 012010 (2017)
42. Ruan, W.-H., Guo, Z.-K., Cai, R.-G., Zhang, Y.-Z.: Int. J. Mod. Phys. A **35**, 2050075 (2020)
43. Maldacena, J.M.: Adv. Theor. Math. Phys. **2**, 231 (1998); Maldacena, J.M.: Int. J. Theor. Phys. **38**, 1113 (1999)
44. Aharony, O., Gubser, S.S., Maldacena, J.M., Ooguri, H., Oz, Y.: Phys. Rep. **323**, 183 (2000)
45. Regge, T., Wheeler, J.A.: Phys. Rev. **108**, 1063 (1957)
46. Carter, B.: Phys. Rev. **174**, 1559 (1968)
47. Frolov, V.P., Kubiznak, D.: Class. Quantum Gravity **25**, 154005 (2008)
48. Bagrov, V.G., Evseevich, A.A., Shapovalov, A.V.: Class. Quantum Gravity **8**, 163 (1991)
49. Frolov, V.P., Krtous, P., Kubiznak, D.: JHEP **0702**, 005 (2007)
50. Son, D.T., Starinets, A.O.: Ann. Rev. Nucl. Part. Sci. **57**, 95 (2007)
51. Chandrasekhar, S., Detweiler, S.L.: Proc. R. Soc. Lond. A **344**, 441 (1975)
52. Kokkotas, K.D., Schmidt, B.G.: Living Rev. Relativ. **2**, 2 (1999)
53. Berti, E., Cardoso, V., Starinets, A.O.: Class. Quantum Gravity **26**, 163001 (2009)
54. Vilenkin, A.: Phys. Lett. B **78**, 301 (1978)

55. Starobinskij, A.A., Churilov, S.M.: Zhurnal Eksperimentalnoi i Teoreticheskoi Fiziki **65**, 3 (1973)
56. Brito, R., Cardoso, V., Pani, P.: Lect. Notes Phys. **906**, 1 (2015)
57. Cardoso, V., Franzin, E., Pani, P.: Phys. Rev. Lett. **116**(17), 171101 (2016)
58. Cardoso, V., Hopper, S., Macedo, C.F.B., Palenzuela, C., Pani, P.: Phys. Rev. D **94**, 084031 (2016)
59. Völkel, S.H., Kokkotas, K.D.: Class. Quantum Gravity **34**, 125006 (2017)
60. Mark, Z., Zimmerman, A., Du, S.M., Chen, Y.: A recipe for echoes from exotic compact objects. arXiv:1706.06155 [gr-qc]
61. Saravani, M., Afshordi, N., Mann, R.B.: Int. J. Mod. Phys. D **23**, 1443007 (2015)
62. Mathur, S.D., Turton, D.: JHEP **01**, 034 (2014)
63. Ferrari, V., Mashhoon, B.: Phys. Rev. D **30**, 295 (1984)
64. Berti, E., Cardoso, V., Will, C.M.: Phys. Rev. D **73**, 064030 (2006)
65. Berti, E., Cardoso, V.: Int. J. Mod. Phys. D **15**, 2209 (2006)
66. Friedman, J.L.: Commun. Math. Phys. **62**, 247 (1978)
67. Moschidis, G.: Commun. Math. Phys. **358**, 437 (2018)
68. Poisson, E., Will, C.: Gravity: Newtonian, Post-Newtonian, Relativistic. Cambridge University Press, Cambridge, UK (2014)
69. Binnington, T., Poisson, E.: Phys. Rev. D **80**, 084018 (2009)
70. Damour, T., Nagar, A.: Phys. Rev. D **80**, 084035 (2009)
71. Porto, R.A.: Fortsch. Phys. **64**, 723 (2016)
72. Cardoso, V., Franzin, E., Maselli, A., Pani, P., Raposo, G.: Phys. Rev. D **95**, 084014 (2017)
73. Maselli, A., Pani, P., Cardoso, V., Abdelsallhin, T., Gualtieri, L., Ferrari, V.: Phys. Rev. Lett. **120**, 081101 (2018)
74. Nollert, H.P.: Class. Quantum Gravity **16**, R159 (1999)
75. Konoplya, R.A., Zhidenko, A.: Rev. Mod. Phys. **83**(3) (2011)
76. Cardoso, V., Pani, P.: Living Rev. Relativ. **22**, 4 (2019)
77. Ferrari, V., Gualtieri, L.: Gen. Relativ. Gravit. **40**, 945 (2008)
78. Cardoso, V., Pani, P.: Nat. Astron. **1**, 586 (2017)
79. Audley, H., et al.: Laser Interferometer Space Antenna. arXiv:1702.00786
80. Banks, T., Fischler, W., Shenker, S.H., Susskind, L.: Phys. Rev. D **55**, 5112 (1997)
81. Susskind, L., Witten, E.: The Holographic bound in anti-de Sitter space. arXiv:hep-th/9805114
82. Susskind, L., Lindesay, J.: An Introduction to Black Holes, Information and the String Theory Revolution: The Holographic Universe. World Scientific Publishing Co. Inc. (2004)
83. Bousso, R.: Rev. Mod. Phys. **74**, 825 (2002)
84. Fischler, W., Susskind, L.: Holography and cosmology. arXiv:hep-th/9806039
85. Page, D.N.: Phys. Rev. Lett. **71**, 3743 (1993)
86. Corley, S., Jacobson, T.: Phys. Rev. D **53**, R6720 (1996)
87. Susskind, L.: J. Math. Phys. **36**, 6377 (1995)
88. Kawai, H., Lewellen, D.C., Tye, S.-H.H.: Nucl. Phys. B **269**, 1 (1986)
89. Bern, Z.: Living Rev. Relativ. **5**, 5 (2002)
90. Mizera, S.: JHEP **1708**, 097 (2017)
91. Bern, Z., Carrasco, J.J.M., Johansson, H.: Phys. Rev. Lett. **105**, 061602 (2010)
92. Jordan, P.: Z. Phys. **93**, 464 (1935); **98**, 759 (1936); **99**, 109 (1936); **102**, 243 (1936); **105**, 114, 229 (1937)
93. Pryce, M.H.L.: Proc. R. Soc. A **165**, 247 (1938)
94. Feynman, R.P., Morinigo, F.P., Wagner, W.G.: Feynman Lectures on Gravitation. CRC Press (2018)
95. Papini, G.: Nuovo Cimento **39**, 716 (1965)
96. Borsten, L., Kim, H., Jurko, B., Macrelli, T., Saemann, C., Wolf, M.: Double Copy from Homotopy Algebras. arXiv:2102.11390
97. Bose, S., Mazumdar, A., Morley, G.W., Ulbricht, H., Toros, M., Paternostro, M., Geraci, A.A., Barker, P.F., Kim, M.S., Milburn, G.: Phys. Rev. Lett. **119**, 240401 (2017)
98. Marletto, C., Vedral, V.: Phys. Rev. Lett. **119**, 240402 (2017)

99. Rovelli, C.: Quantum Gravity. Cambridge Monographs on Mathematical Physics, Cambridge University Press, Cambridge (2004); Rovelli, C., Vidotto, F.: Covariant Loop Quantum Gravity. Cambridge University Press (2014); Smolin, L.: arXiv:0507235 (2005)
100. Polchinski, J.: An Introduction to the Bosonic String. Cambridge University Press (2011). Kiritsis, E.: String Theory in a Nutshell. Princeton University Press (2019); Dine, M.: Supersymmetry and String Theory: Beyond the Standard Model. Cambridge University Press (2016); Green, M.B., Schwarz, J.H., Witten, E.: Superstring Theory. Cambridge University Press (2012)

Conclusions 9

The general theory of relativity has met with remarkable success in explaining phenomena where the values of physical quantities are determined individually and independently. Then quantum physics came along and proved equally successful in a domain of nature where conjugate observables are known only within confines determined by uncertainty relations.

While quantum field theories have been developed, with difficulties, for the weak, electromagnetic, and strong forces, an accepted, comprehensive quantum field theory of gravitation does not exist. This is mainly due to the fact that space–time itself must be quantized in the case of gravitation.

Between the scale of events where general relativity is well tested and the Planck scale, where both gravitation and quantum effects become equally important, there is a large area where general relativity and quantum physics meet and coexist. The hope is that their interaction in this twilight zone show precursory aspects of their innermost behaviour. Exploring this area has not been easy, despite the enormous progress of instrumentation. It is an area where general relativity maintains its overwhelmingly classical behaviour, but may interact, in observable ways, with quantum systems. It is therefore expected that a quasi-classical theory would be adequate. A theory with these properties, referred to as external field approximation, has not been available until the 1980s and later largely because of the mathematically complex structure of Einstein's theory that involves potentials with ten components rather than only four as in electromagnetism. In EFA, gravity is described by classical equations when it interacts with quantum particles. It is patterned after a similar theory in electrodynamics where radiation processes are greatly enhanced when incident photons are replaced by external electromagnetic fields [1]. A consistent, external field approximation now exists. It has its roots in quantum wave equations and in Berry phase and can be used to calculate effects that, though small, may still be of interest in astrophysics. Without these effects a (low energy) gravitational astrophysics

G. Lambiase and G. Papini, *The Interaction of Spin with Gravity in Particle Physics*, Lecture Notes in Physics 993, https://doi.org/10.1007/978-3-030-84771-5_9

may never exist. This approximation has been developed in Chap. 1 and applied to particular physical problems in Chaps. 2 and 3 and to neutrinos. It is then shown in Chap. 6 that EFA enables the existence of radiative effects otherwise kinematically forbidden, and of spin currents. The latter development stems from the possibility to separate the flow of charge from that of spin angular momentum. Direct analysis of the spin current tensor shows that the presence of gravity invalidates the separate conservation of angular momentum and spin and allows the continual interchange of spin and orbital angular momentum as in electromagnetism.

Related to spin and EFA is the possibility to generate vortices by applying gravitational fields to ensembles of identical particles. The subject is actively investigated experimentally and theoretically.

Among the first genuinely quantum systems met in this investigation are charged and neutral superfluids. They exhibit quantization on a macroscopic scale, which, for the present purposes, seems to be an ideal property to have in both classical and quantum contexts. The wave function of a superconductor describes a large number of coupled electrons in the same quantum state and acquires quasi-classical properties. Curvature appears explicitly in a quantum environment and the calculation of Berry phase for closed loops leads to quantization. Gravitation and fictitious gravitational forces produce quantum phases.

Observations in EFA amount to measurements of phase differences. These can be carried out by means of interferometers of which several have been used [2,3], are in present use, or under construction [4–10]. Interferometry has indeed been the main subject of Chap. 3.

Spin is also a quintessential quantum object of interest in this work. Among important theoretical developments, due to Mashhoon in the 1990s [11], was indeed the realization that a coupling between gravitation and spin should exist in nature. The effect has been singled out and observed for photons [12] and neutrons [13,14]. An extension of spin-rotation coupling to compound spin systems has been given in Chap. 2.

The coupling of gravity to spin is responsible for the oscillations of the helicity and chirality of fermions. A discussion of this subject has been given in Chap. 2 and has been studied in some detail for neutrinos in Chap. 4. Other aspects of the interaction, particularly at the astrophysical level, have been examined in Chap. 5.

Additional aspects of the particle–gravity interaction have been considered in Chap. 6. They include the occurrence of quasiparticles as quanta of the field $K_\lambda(x)$ and of transformations like (1.37) that render the ground state of the system space–time dependent and lead to the phenomenon of symmetry breaking. The application of Stokes theorem to Eqs. (1.37) and (1.38) leads, by transportation along a closed path Γ, to a change of ϕ_0 given by

$$\delta\phi \sim \int_\Sigma df^{\beta\delta} R_{\mu\nu\beta\delta} J^{\mu\nu} \phi_0 = \int_\Sigma df^{\beta\delta} (K_{\beta,\delta} - K_{\delta,\beta}) .$$

The rotational invariance of ϕ_0 is therefore violated by

$$[\partial z_\beta, \partial z_\delta]\Phi_g = R_{\mu\nu\beta\delta} J^{\mu\nu} ,$$

which obviously vanishes in flat space–time, or when $J^{\mu\nu} = 0$. Because of the structure of Eq. (1.41), this result can be extended to higher orders of $\gamma_{\mu\nu}$.

As the sensitivity of instrumentation advances, the horizon widens and limits on violations of conservation laws also become more stringent. This is the case for discrete symmetries where spin plays a fundamental role. Spin and gravitation endow space–time with a measure of chirality. In a purely geometrical context, this could be represented by torsion. Axions also account for chirality in the universe and are considered leading candidates for dark matter. Their discovery would provide a solution to two of the major problems of modern physics, the violation of strong CP invariance and the origin of dark matter. The search for axions and their interaction with spin and matter are the subject of Chap. 7 and passes through the development of appropriate forms of the Bargmann–Michel–Telegdi equation [15] and the formulation of a theory of axion electrodynamics by Ni [16] and Sikivie [17]. Several experimental searches are under way [18, 19].

Research subjects of relevant current interest like gravitational memory, QNMs, holography, double copy, and KLT relations are briefly mentioned in Chap. 8.

What has been discussed in the previous chapters should not be construed as an attempt to build a quantum theory of gravitation from its low-energy limit. It would be wrong as building QED from classical electromagnetism has proven. The objective has been to rather show that the stage defined by macroscopic scales on one side and Planck length on the other is not empty. The dimensions of his stage are so vast that plots and facts agitated on it will probably persist even after the advent of a successful theory of quantum gravity. Gravitational astrophysics need not consist only of high-energy phenomena.

The problem to bring together gravitation and quantum physics even semiclassically is objectively difficult. However, what in the 1980s looked like a hopeless search for a role of gravity at the quantum level, before the Planck scale threshold, is proving fruitful in unexpected ways in the study of gravitation, its invariance limits, symmetry violations, astrophysics, the structure of space–time, and the origin of dark matter and dark energy.

References

1. Jauch, J.M., Rohrlich, F.: The Theory of Photons and Electrons. Springer, New York (1976)
2. Werner, S.A., Staudenman, J.-L., Colella, R.: Phys. Rev. Lett. **42**, 1103 (1979)
3. Werner, S.A., Kaiser, H.: Neutron interferometry—macroscopic manifestation of quantum mechanics. In: Audretsch, J., DeSabbata, V. (eds.) Quantum Mechanics in Curved Space-Time. Plenum Press, New York (1990)
4. Riehle, F., Kister, Th., Witte, A., Helmcke, J., Bordé, Ch.J.: Phys. Rev. Lett. **67**, 177 (1991); Bordé, Ch.J.: Phys. Lett. A **140**, 10 (1989); Bordé, Ch.J., et al.: Phys. Lett. A **188**, 187 (1997); Bordé, Ch.J.: In: Berman, P. (ed.) Atom Interferometry. Academic Press, London (1997)
5. Salvi, L., Poli, N., Voletic, V., Tino, G.M.: Phys. Rev. Lett. **120**, 033601 (2018); Lu, L., Poli, N., Salvi, L., Tino, G.M.: Phys. Rev. Lett. **119**, 263601 (2017); Rosi, G., Cacciapuoti, L., Sorrentino, F., Menchetti, M., Prevedelli, M., Tino, G.M.: Phys. Rev. Lett. **114**, 013001 (2015)
6. Duan, X.-C., Den, X.-B., Zhou, M.-K., Ke-Zhang, Xu, W.-L., Shao, C.-G., Luo, J., Hu, Z.-K.: Phys. Rev. Lett. **117**, 023001 (2016)

7. Howl, R., Hackermüller, L., Bruschi, D.E., Fuentes, I.: Adv. Phys.: X **3**(1), 1383184 (2018)
8. Abele, H., Jenke, T., Leeb, H., Schmiedmayer, J.: Phys. Rev. D **81**, 065019 (2010)
9. Nesvizhevsky, V.V., Börner, H.G., Petukhov, A.K., Abele, H., BaeBler, S., Ruess, F.J., Stöferle, T., Westphal, A., Gagarski, A.M., Petrov, G.A., Srelkov, A.V.: Nature **415**, 297 (2002)
10. Jenke, T., Geltenbort, P., Lemmel, H., Abele, H.: Nat. Phys. **7**, 468 (2011)
11. Mashhoon, B.: Phys. Rev. Lett. **61**, 2639 (1988); Phys. Lett. A **139**, 103 (1989); **143**, 176 (1990); **145**, 147 (1990); Phys. Rev. Lett. **68**, 3812 (1992)
12. Ashby, N.: Living Rev. Relativ. **6**, 1 (2003)
13. Demirel, B., Sponar, S., Hasegawa, Y.: New J. Phys. **17**, 023065 (2015)
14. Danner, A., Demirel, B., Sponar, S., Hasegawa, Y.: J. Phys. Commun. **3**, 035001 (2019)
15. Balakin, A.B., Popov, V.A.: Phys. Rev. D **92**, 105025 (2015)
16. Ni, W.-T.: Phys. Rev. Lett. **38**, 301 (1977)
17. Sikivie, P.: Phys. Rev. Lett. **51**, 1415 (1983)
18. See, e.g., Chang, S.P., Haciomeruglu, S., Kim, O., Lee, S., Park, S., Semertzidis, Y.K.: XVII International Workshop on Polarizes Sources, Targets and Polarimetry, Kaist, South Korea, 16–20 Oct 2017
19. Mohanty, S., Mukhopadhyay, B., Prasanna, A.R.: arXiv:hep-ph/0204257 [hep-ph]

Natural Units and Conventions

A

Planck units are used throughout this work, unless explicitly stated.

$$k_B = c = \hbar = G = 1\,, \tag{A.1}$$

with

$$\begin{aligned} k_B &= 1.3807\ 10^{-16} erg K^{-1}\,, \\ \hbar &= 1.0546\ 10^{-27} cm^2 g s^{-1}\,, \\ c &= 2.9979\ 10^{10} cm s^{-1}\,, \\ G &= 6.6720\ 10^{-8} cm^3 g^{-1} s^{-2}\,. \end{aligned}$$

The conversion to physics units follows by using

$$\begin{aligned} l_p &= (G\hbar/c^3)^{1/2}\,, \\ t_p &= (G\hbar/c^5)^{1/2}\,, \\ m_p &= (\hbar/G)^{1/2}\,. \end{aligned}$$

Equation (A.1) implies

$$[energy] = [mass] = [temperature] = [length]^{-1} = [time]^{-1}\,.$$

Moreover, the following conversion relations hold:

$$1m = 5.07 \times 10^{15} GeV^{-1}, \tag{A.2}$$
$$1sec = 1.52 \times 10^{24} GeV^{-1}, \tag{A.3}$$
$$1K = 9 \times 10^{-5} eV, \tag{A.4}$$
$$1cm^3 = 1.25 \times 10^{14} eV^{-3}, \tag{A.5}$$
$$1MeV = 4.6 \times 10^{31} cm/sec^2, \tag{A.6}$$
$$1Gauss = 10^{-4}T = 7 \times 10^{-20} GeV^2, \tag{A.7}$$
$$1gr/cm^3 = 4.5 \times 10^{18} eV^4. \tag{A.8}$$

G. Lambiase and G. Papini, *The Interaction of Spin with Gravity in Particle Physics*, Lecture Notes in Physics 993, https://doi.org/10.1007/978-3-030-84771-5_A

For the astronomical units, one has

$$1\,pc = 3.0856\ 10^{18} cm.$$

The Hubble constant is

$$H_0 = 100\,h K m s^{-1} Mpc^{-1}\,, \quad 0.4 \leq h \leq 1\,, \tag{A.9}$$

while the Hubble time is

$$\frac{1}{H_0} = 3.0856\ 10^{17} h^{-1} s\,, \tag{A.10}$$

and the Hubble distance is

$$c H_0 = 29997.9\,h\,Mpc\,. \tag{A.11}$$

We adopt the signature is $(+, -, -, -)$. For the Riemann tensor, we define

$$R^{\alpha}{}_{\beta\mu\nu} = \partial_\mu \Gamma^{\alpha}_{\beta\nu} + \cdots\,, \tag{A.12}$$

while for the Ricci tensor we use the convention

$$R_{\mu\nu} = R^{\sigma}{}_{\mu\sigma\nu}\,. \tag{A.13}$$

The affinity connections are the usual Christoffel symbols of the metric

$$\Gamma^{\mu}_{\alpha\beta} = \frac{1}{2} g^{\mu\sigma} \left(g_{\alpha\sigma,\beta} + g_{\beta\sigma,\alpha} - g_{\alpha\beta,\sigma} \right). \tag{A.14}$$

The trace is defined as

$$T = T^{\mu}_{\ \mu} = g^{\mu\nu} T_{\mu\nu}\,. \tag{A.15}$$

The ordinary derivative of a tensor $\mathfrak{T}$ is indicated as

$$\frac{\partial \mathfrak{T}}{\partial x^\mu} \equiv \partial_\mu \mathfrak{T} \equiv \mathfrak{T}_{,\mu}\,. \tag{A.16}$$

Dirac Matrices

B

Some properties of the Dirac matrices are provided below. They are determined by the Clifford algebra

$$\{\gamma^\mu, \gamma^\nu\} = 2\eta^{\mu\nu} I\,. \tag{B.1}$$

From the trace properties, $TrAB = TrBA$, $Tr([A, B]) = 0$ for any A and B, and $Tr(ABC) = Tr(CAB) = Tr(BCA)$ (cyclic property), one gets

$$Tr(\gamma^\mu\gamma^\nu) = 4\eta^{\mu\nu}\,. \tag{B.2}$$

The trace of the product of odd number of Dirac matrices vanishes

$$Tr\gamma^\mu = 0\,, \quad Tr\left(\gamma^\mu\gamma^\nu\gamma^\rho\right) = 0\,, \quad Tr\left(\gamma^{\mu_1}\gamma^{\mu_2}\ldots\gamma^{\mu_{2n+1}}\right) = 0\,. \tag{B.3}$$

Moreover, one has

$$Tr\left(\gamma^\kappa\gamma^\lambda\gamma^\mu\gamma^\nu\right) = 4\left(\eta^{\kappa\lambda}\eta^{\mu\nu} - \eta^{\kappa\mu}\eta^{\lambda\nu} + \eta^{\kappa\nu}\eta^{\lambda\mu}\right),$$

$$\begin{aligned}
Tr\left(\gamma^\kappa\gamma^\lambda\gamma^\mu\gamma^\nu\gamma^\rho\gamma^\sigma\right) = {} & 4\eta^{\kappa\lambda}\left(\eta^{\mu\nu}\eta^{\rho\sigma} - \eta^{\mu\rho}\eta^{\nu\sigma} + \eta^{\mu\sigma}\eta^{\nu\rho}\right) - \\
& - 4\eta^{\kappa\mu}\left(\eta^{\lambda\nu}\eta^{\rho\sigma} - \eta^{\lambda\rho}\eta^{\nu\sigma} + \eta^{\lambda\sigma}\eta^{\nu\rho}\right) + \\
& + 4\eta^{\kappa\nu}\left(\eta^{\lambda\mu}\eta^{\rho\sigma} - \eta^{\lambda\rho}\eta^{\nu\sigma} + \eta^{\lambda\sigma}\eta^{\nu\rho}\right) - \\
& - 4\eta^{\kappa\nu}\left(\eta^{\lambda\mu}\eta^{\rho\sigma} - \eta^{\lambda\rho}\eta^{\mu\sigma} + \eta^{\lambda\sigma}\eta^{\nu\rho}\right) - \\
& - 4\eta^{\kappa\rho}\left(\eta^{\lambda\mu}\eta^{\nu\sigma} - \eta^{\lambda\nu}\eta^{\mu\sigma} + \eta^{\lambda\rho}\eta^{\mu\nu}\right) + \\
& + 4\eta^{\kappa\sigma}\left(\eta^{\lambda\mu}\eta^{\nu\rho} - \eta^{\lambda\nu}\eta^{\mu\rho} + \eta^{\lambda\rho}\eta^{\mu\nu}\right).
\end{aligned}$$

G. Lambiase and G. Papini, *The Interaction of Spin with Gravity in Particle Physics*, Lecture Notes in Physics 993, https://doi.org/10.1007/978-3-030-84771-5_B

In general, for an even number of Dirac matrices, one has the recursive formula

$$Tr\left(\gamma^{\mu_1}\gamma^{\mu_2}\ldots\gamma^{\mu_n}\right) = \sum_{k=2}^{n}(-1)^k \eta^{\nu_1\nu_k}\, Tr\left(\gamma^{\nu_2}\ldots \underbrace{\gamma^{\nu_k}}_{remove}\ldots\gamma^{\nu_n}\right), \tag{B.4}$$

with the k-th matrix γ^{μ_k} to be removed. Moreover, $\gamma^{\mu\,\dagger} = \gamma^0\gamma^\mu\gamma^0$.

Besides the Dirac matrices, one also defines the γ^5 matrix

$$\gamma^5 = i\gamma^0\gamma^1\gamma^2\gamma^3, \tag{B.5}$$

with the properties

$$\gamma^{5\dagger} = \gamma^5, \quad \left(\gamma^5\right)^2 = I, \quad \{\gamma^\mu, \gamma^5\} = 0. \tag{B.6}$$

For the trace properties, one finds

$$Tr\gamma^5 = 0, \quad Tr\left(\gamma^5\gamma^\mu\right) = 0, \quad Tr\left(\gamma^5\gamma^\mu\gamma^\nu\right) = 0, \quad Tr\left(\gamma^5\gamma^\mu\gamma^\nu\gamma^\lambda\right) = 0, \tag{B.7}$$

and

$$Tr\left(\gamma^5\gamma^\kappa\gamma^\lambda\gamma^\mu\gamma^\nu\right) = -4i\varepsilon^{\kappa\lambda\mu\nu}, \qquad Tr\left(\gamma^5\gamma^{\mu_1}\ldots\gamma^{\mu_{2n+1}}\right) = 0, \tag{B.8}$$

$$\gamma^5\gamma^\lambda\gamma^\mu\gamma^\nu = \eta^{\lambda\mu}\gamma^5\gamma^\nu - \eta^{\lambda\nu}\gamma^5\gamma^\mu + \eta^{\mu\nu}\gamma^5\gamma^\lambda - i\varepsilon^{\lambda\mu\nu\rho}\gamma_\rho. \tag{B.9}$$

In the Dirac representation, the γ^μ and γ^5 assume the form

$$\gamma^0 = \begin{pmatrix} I & 0 \\ 0 & -I \end{pmatrix} \quad \gamma^i = \begin{pmatrix} 0 & \sigma^i \\ -\sigma^i & 0 \end{pmatrix}, \quad \gamma^5 = \begin{pmatrix} 0 & I \\ I & 0 \end{pmatrix}. \tag{B.10}$$

Finally, the Pauli matrices are

$$\sigma^1 = \begin{pmatrix} 0 & 1 \\ 1 & 0 \end{pmatrix}, \quad \sigma^2 = \begin{pmatrix} 0 & -i \\ i & 0 \end{pmatrix}, \quad \sigma^3 = \begin{pmatrix} 1 & 0 \\ 0 & -1 \end{pmatrix} \tag{B.11}$$

and satisfy the relation

$$\sigma^i\sigma^j = \delta^{ij} I + i\,\varepsilon^{ijk}\sigma^k. \tag{B.12}$$

Neutrino Oscillations in Flat and Curved Space–Time

C

The flavour eigenstate $|\nu_\alpha\rangle$, with $\alpha = e, \mu, \tau$, and mass eigenstate $|\nu_i\rangle$, with $i = 1, 2, 3$, are related by

$$|\nu_\alpha\rangle = \sum_i U^*_{\alpha i}\,|\nu_i\rangle\,, \tag{C.1}$$

where U is the 3×3 unitary matrix (the mixing matrix). Taking the neutrino wave function as a plane wave, its propagation from source S with coordinates $(t_S, \mathbf{x}_S)$ to the detector D with coordinates $(t_D, \mathbf{x}_D)$ is given by

$$|\nu_i(t_D, \mathbf{x}_D)\rangle = e^{-i\Phi_i}\,|\nu_i(t_S, \mathbf{x}_S)\rangle\,, \tag{C.2}$$

while the probability of the change in neutrino flavour from $\nu_\alpha \to \nu_\beta$ at the detection point D is given by

$$P_{\alpha\beta} \equiv \left|\langle\nu_\beta|\nu_\alpha(t_D, \mathbf{x}_D)\rangle\right|^2 = \sum_{i,j} U_{\beta i} U^*_{\beta j} U_{\alpha j} U^*_{\alpha i}\, e^{-i(\Phi_i - \Phi_j)}\,, \tag{C.3}$$

if the neutrinos are produced initially in the flavour eigenstate $|\nu_\alpha\rangle$ at S. In flat space–time, the phase is given by

$$\Phi_i = E_i(t_D - t_S) - \mathbf{p}_i \cdot (\mathbf{x}_D - \mathbf{x}_S)\,. \tag{C.4}$$

Assuming that all the mass eigenstates in a flavour eigenstate initially produced at the source have equal momentum or energy [1,2], and that $(t_D - t_S) \simeq |\mathbf{x}_D - \mathbf{x}_S|$ for relativistic neutrinos ($E_i \gg m_i$), one gets

$$\Delta\Phi_{ij} \equiv \Phi_i - \Phi_j \simeq \frac{\Delta m^2_{ij}}{2E_0}\,|\mathbf{x}_D - \mathbf{x}_S|, \tag{C.5}$$

G. Lambiase and G. Papini, *The Interaction of Spin with Gravity in Particle Physics*, Lecture Notes in Physics 993, https://doi.org/10.1007/978-3-030-84771-5_C

where $\Delta m_{ij}^2 \equiv m_i^2 - m_j^2$. E_0 is the average energy of the relativistic neutrinos produced at S, while, considering only two flavours of neutrinos, the oscillation formula becomes

$$P_{e\mu} = \sin^2 2\alpha \, \sin^2\left(\frac{\Delta m_{12}^2 L}{4E_0}\right) \tag{C.6}$$

where $L = |\mathbf{x}_D - \mathbf{x}_S|$. The angle α parametrizes the 2×2 matrix

$$U \equiv \begin{bmatrix} \cos\alpha & \sin\alpha \\ -\sin\alpha & \cos\alpha \end{bmatrix}, \tag{C.7}$$

relating the flavour and mass eigenbases for this case.

In a curved space–time, the phase (C.4) is replaced by [3].

$$\Phi_i = \int_S^D p_\mu^{(i)} \, dx^\mu , \tag{C.8}$$

where $p_\mu^{(i)}$ is the canonical conjugate momentum that refers to the ith neutrino mass eigenstate

$$p_\mu^{(i)} = m_i \, g_{\mu\nu} \frac{dx^\mu}{ds} . \tag{C.9}$$

Here, $g_{\mu\nu}$ is the metric tensor and ds is the line element along the neutrino trajectory. For a gravitational field of a static spherically symmetric object described by the Schwarzschild metric [3,4], one finds

$$\Phi_j = \int_S^D \left(E_j(r)\, dt - p_j(r)\, dr - J_j(r)\, d\phi\right) , \tag{C.10}$$

where $E_j(r) \equiv p_t^{(j)}$, $p_j(r) \equiv -p_r^{(j)}$ and $J_j(r) \equiv -p_\phi^{(j)}$.

To obtain the phase, one assumes that the neutrinos follow the classical trajectory from the source to the detector. In the quantum framework, the use of classical trajectories seems unjustified. However, such an approximation is justified for a relativistic quantum particle in the regime of sufficiently weak gravitational field (see [5] and references therein). Assuming $GM \ll r$ (weak gravity limit) and $b \ll r_{S,D}$, one finds [4]

$$\Phi_j \simeq \frac{m_j^2}{2E_0}(r_S + r_D)\left(1 - \frac{b^2}{2r_S r_D} + \frac{2GM}{r_S + r_D}\right) . \tag{C.11}$$

The phase difference $\Delta\Phi_{jk}$ (C.11) depends on Δm_{jk}^2, as well as, in curved space–time, on the gravitational lensing through different paths.

Dirac Hamiltonians in the Low-Energy Approximation

D

Ordinary units are used in this appendix. Equation (6.53) can now be used to derive the Hamiltonian in general coordinates

$$
\begin{aligned}
i\hbar c\,\nabla_0\psi'(x) = (g^{00}(x))^{-1}\Big[c\,\gamma^0(x)\gamma^j(x)\left(-i\hbar\nabla_j\right) + mc^2\gamma^0(x) \\
+c\,\gamma^0(x)\gamma^\mu(x)(\nabla_\mu\Theta)\Big]\psi'(x) \;=\; H\psi'(x),
\end{aligned}
\tag{D.1}
$$

where

$$
g^{00}(x) = \left(e^0{}_{\hat 0}\right)^2\eta^{\hat 0\hat 0} \;=\; \left(1+\frac{\mathbf{a}\cdot\mathbf{x}}{c^2}\right)^{-2}, \tag{D.2}
$$

$$
\gamma^0(x) = e^0{}_{\hat 0}\gamma^{\hat 0} \;=\; \left(1+\frac{\mathbf{a}\cdot\mathbf{x}}{c^2}\right)^{-1}\beta, \tag{D.3}
$$

$$
\begin{aligned}
\gamma^0(x)\gamma^j(x) &= e^0{}_{\hat 0}\left(\gamma^{\hat 0}\gamma^{\hat j} + e^j{}_{\hat 0}\right) \\
&= \left(1+\frac{\mathbf{a}\cdot\mathbf{x}}{c^2}\right)^{-2}\left[\left(1+\frac{\mathbf{a}\cdot\mathbf{x}}{c^2}\right)\alpha^{\hat j} - \frac{1}{c}\epsilon^{jkl}\,\omega_k\,x_l\right].
\end{aligned}
\tag{D.4}
$$

Since Φ_G is correct only to first order, this is also a constraint on the validity of (D.1). In what follows, terms of higher order in the metric deviation will be dropped. Explicit evaluation of $\nabla_\mu\Theta$ shows that

$$
(\nabla_\mu\Theta) = \nabla_\mu(\Phi_{EM}+\Phi_S+\Phi_G) \;=\; \frac{e}{c}A_\mu + \hbar\Gamma_\mu + (\nabla_\mu\Phi_G), \tag{D.5}
$$

where

$$
\Gamma_0 = -\frac{i}{2c^2}(\mathbf{a}\cdot\boldsymbol{\alpha}) - \frac{1}{2c}\boldsymbol{\omega}\cdot\boldsymbol{\sigma}, \tag{D.6}
$$

$$
\Gamma_j = 0, \tag{D.7}
$$

G. Lambiase and G. Papini, *The Interaction of Spin with Gravity in Particle Physics*, Lecture Notes in Physics 993, https://doi.org/10.1007/978-3-030-84771-5_D

and

$$(\nabla_\mu \Phi_G) = \frac{1}{2}\gamma_{\alpha\mu}(x)p^\alpha - \frac{1}{2}\int_X^x dz^\lambda(\gamma_{\mu\lambda,\beta}(z) - \gamma_{\beta\lambda,\mu}(z))p^\beta, \tag{D.8}$$

where p^μ is the momentum eigenvalue of the free particle.

It follows that, to first order in $\mathbf{a}$ and $\boldsymbol{\omega}$, the Dirac Hamiltonian in the general coordinate frame is

$$H \approx c(\boldsymbol{\alpha}\cdot\mathbf{p}) + mc^2\beta + V(\mathbf{x}), \tag{D.9}$$

where

$$\begin{aligned} V(\mathbf{x}) = {} & \frac{1}{c}(\mathbf{a}\cdot\mathbf{x})(\boldsymbol{\alpha}\cdot\mathbf{p}) + m(\mathbf{a}\cdot\mathbf{x})\beta - \boldsymbol{\omega}\cdot(\mathbf{L}+\mathbf{S}) - \frac{i\hbar}{2c}(\mathbf{a}\cdot\boldsymbol{\alpha}) \\ & - e\left(1 + \frac{\mathbf{a}\cdot\mathbf{x}}{c^2}\right)(\boldsymbol{\alpha}\cdot\mathbf{A}) + \frac{e}{c}\,\boldsymbol{\omega}\cdot(\mathbf{x}\times\mathbf{A}) + e\,\varphi \\ & + c\,\boldsymbol{\alpha}\cdot(\boldsymbol{\nabla}\Phi_G) + c\,(\nabla_0\Phi_G), \end{aligned} \tag{D.10}$$

the $\boldsymbol{\alpha}$, β, $\boldsymbol{\sigma}$ matrices are those of Minkowski space, and $\mathbf{L} = \mathbf{x}\times\mathbf{p}$ and $\mathbf{S} = \hbar\boldsymbol{\sigma}/2$ are the orbital and spin angular momenta, respectively.

D.1 Low-Energy Approximation

Although the Dirac Hamiltonian as described by (D.9) and (D.10) is useful as is, there are benefits in considering approximations which emphasize the low-energy limit in a particle's range of motion.

According to the Foldy–Wouthuysen transformation procedure [6], it is possible to write the Dirac Hamiltonian in the form

$$H = mc^2\beta + \mathcal{O} + \mathcal{E}, \tag{D.11}$$

where the "odd" and "even" operators $\mathcal{O}$ and $\mathcal{E}$, respectively, satisfy $\{\mathcal{O}, \beta\} = [\mathcal{E}, \beta] = 0$. The anomalous magnetic moment can be introduced by adding to V the term

$$\frac{\kappa e\hbar}{2mc}\sigma^{\mu\nu}F_{\mu\nu} = \frac{\kappa e\hbar}{2mc}(i\boldsymbol{\alpha}\cdot\mathbf{E} - \boldsymbol{\sigma}\cdot\mathbf{B}), \tag{D.12}$$

with $\kappa \equiv (g-2)/2$, by means of the substitution

$$mc^2\beta \;\rightarrow\; \beta\left[mc^2 + \frac{\kappa e\hbar}{2mc}(i\boldsymbol{\alpha}\cdot\mathbf{E} - \boldsymbol{\sigma}\cdot\mathbf{B})\right]. \tag{D.13}$$

Then, by comparing (D.9) and (D.10), one finds

$$\mathcal{O} = c\boldsymbol{\alpha} \cdot \left[\left(1 + \frac{\mathbf{a}\cdot\mathbf{x}}{c^2}\right)\boldsymbol{\pi} + (\nabla\Phi_{\mathrm{G}}) - \frac{i\kappa e\hbar}{2mc^2}\beta\mathbf{E} - \frac{i\hbar}{2c^2}\mathbf{a}\right], \tag{D.14}$$

$$\begin{aligned}\mathcal{E} &= \left[m(\mathbf{a}\cdot\mathbf{x}) - \frac{\kappa e\hbar}{2mc}\left(1 + \frac{\mathbf{a}\cdot\mathbf{x}}{c^2}\right)(\boldsymbol{\sigma}\cdot\mathbf{B})\right]\beta \\ &\quad -\boldsymbol{\omega}\cdot(\mathbf{x}\times\boldsymbol{\pi}) - \boldsymbol{\omega}\cdot\left(\frac{\hbar}{2}\boldsymbol{\sigma}\right) + e\varphi + c(\nabla_0\Phi_{\mathrm{G}}),\end{aligned} \tag{D.15}$$

where $\boldsymbol{\pi} = \mathbf{p} - e\mathbf{A}/c$.

Following the procedure given by Bjorken and Drell [7], the transformed Hamiltonian is represented by a series expansion of S, according to a unitary transformation

$$\begin{aligned}H' &= UHU^{-1} \\ &\approx H + i[S, H] - \frac{1}{2}[S, [S, H]] - \frac{i}{6}[S, [S, [S, H]]] \\ &\quad + \frac{mc^2}{24}[S, [S, [S, [S, \beta]]]] - \hbar\dot{S} - \frac{i\hbar}{2}[S, \dot{S}],\end{aligned} \tag{D.16}$$

and $S = \mathrm{O}(1/m)$ is the Hermitean exponent of a unitary transformation operator $U \equiv \exp(iS)$. By three successive applications of (D.16) for the choice

$$S \equiv S_{\mathrm{FW}} = -\frac{i}{2mc^2}\beta\mathcal{O}, \tag{D.17}$$

the transformed Hamiltonian becomes

$$\begin{aligned}H_{\mathrm{FW}} &= mc^2\beta + \mathcal{E}' \\ &= \beta\left(mc^2 + \frac{1}{2mc^2}\mathcal{O}^2 - \frac{1}{8m^3c^6}\mathcal{O}^4\right) + \mathcal{E} \\ &\quad - \frac{1}{8m^2c^4}[\mathcal{O}, [\mathcal{O}, \mathcal{E}]] - \frac{i\hbar}{8m^2c^4}[\mathcal{O}, \dot{\mathcal{O}}].\end{aligned} \tag{D.18}$$

To determine the gravitational corrections in (D.18), it is necessary to isolate the external electromagnetic potentials within the definition of odd and even operators. This implies that $\mathcal{O} \equiv \mathcal{O}_0 + \mathcal{O}_1$ and $\mathcal{E} \equiv \mathcal{E}_0 + \mathcal{E}_1$, where

$$\mathcal{O}_0 = c\boldsymbol{\alpha}\cdot\boldsymbol{\pi}, \tag{D.19}$$

$$\mathcal{E}_0 = e\varphi\,. \tag{D.20}$$

Therefore,

$$\mathcal{O}_1 = c\,\boldsymbol{\alpha}\cdot\left[\left(\frac{\mathbf{a}\cdot\mathbf{x}}{c^2}\right)\boldsymbol{\pi} + (\nabla\Phi_{\mathrm{G}}) - \frac{i\kappa e\hbar}{2mc^2}\beta\mathbf{E} - \frac{i\hbar}{2c^2}\mathbf{a}\right], \tag{D.21}$$

$$\mathcal{E}_1 = \left[m(\mathbf{a}\cdot\mathbf{x}) - \frac{\kappa e\hbar}{2mc}\left(1+\frac{\mathbf{a}\cdot\mathbf{x}}{c^2}\right)(\boldsymbol{\sigma}\cdot\mathbf{B})\right]\beta \tag{D.22}$$
$$- \boldsymbol{\omega}\cdot(\mathbf{x}\times\boldsymbol{\pi}) - \boldsymbol{\omega}\cdot\left(\frac{\hbar}{\mathbf{2}}\boldsymbol{\sigma}\right) + \mathbf{c}(\nabla_0\mathbf{8}_{\mathrm{G}}).$$

Neglecting the $\mathcal{O}^4$ contribution and considering only terms up to first order in $\mathbf{a}$, $\boldsymbol{\omega}$, and $1/m^2$, one gets

$$\mathcal{O}^2 = \mathcal{O}_0^2 + \mathcal{O}_1^2 + \{\mathcal{O}_0, \mathcal{O}_1\}, \tag{D.23}$$

$$[\mathcal{O},[\mathcal{O},\mathcal{E}]] \approx [\mathcal{O}_0,[\mathcal{O}_0,\mathcal{E}_0]] + [\mathcal{O}_0,[\mathcal{O}_1,\mathcal{E}_0]] + [\mathcal{O}_0,[\mathcal{O}_0,\mathcal{E}_1]] + \tag{D.24}$$
$$+ [\mathcal{O}_1,[\mathcal{O}_0,\mathcal{E}_0]],$$

$$[\mathcal{O},\dot{\mathcal{O}}] \approx [\mathcal{O}_0,\dot{\mathcal{O}}_0] + [\mathcal{O}_0,\dot{\mathcal{O}}_1] + [\mathcal{O}_1,\dot{\mathcal{O}}_0]. \tag{D.25}$$

From the zeroth-order terms in (D.18), it can be shown that [7]

$$H_{\mathrm{FW}(0)} = mc^2\beta + \frac{1}{2mc^2}\beta\mathcal{O}_0^2 + e\varphi - \frac{1}{8m^2c^4}[\mathcal{O}_0,[\mathcal{O}_0,\mathcal{E}_0]] - \tag{D.26}$$
$$-\frac{i\hbar}{8m^2c^4}[\mathcal{O}_0,\dot{\mathcal{O}}_0]$$
$$= mc^2\beta + \left[\frac{1}{2m}\pi^2 - \frac{e\hbar}{2mc}\boldsymbol{\sigma}\cdot\mathbf{B}\right]\beta - \frac{e\hbar}{4m^2c^2}\boldsymbol{\sigma}\cdot(\mathbf{E}\times\boldsymbol{\pi})$$
$$- \frac{e\hbar^2}{8m^2c^2}[(\nabla\cdot\mathbf{E}) + i\boldsymbol{\sigma}\cdot(\nabla\times\mathbf{E})] + e\varphi,$$

where the third term coupled to β is the magnetic dipole energy, and the following term is the spin–orbit energy.

Neglecting the time-dependent contributions from (D.25) and considering only those terms up to second order in $\boldsymbol{\pi}$, it follows that

$$\mathcal{O}_1^2 = -\frac{i\kappa e\hbar}{m}\beta\left[\left(\frac{\mathbf{a}\cdot\mathbf{x}}{c^2}\right)(\mathbf{E}\cdot\boldsymbol{\pi}) + \mathbf{E}\cdot(\nabla\Phi_{\mathrm{G}})\right] - \frac{\kappa e\hbar^2}{2mc^2}\beta(\mathbf{a}\cdot\mathbf{E}) \tag{D.27}$$
$$-\frac{\kappa e\hbar^2}{2m}\left(\frac{\mathbf{a}\cdot\mathbf{x}}{c^2}\right)\beta(\nabla\cdot\mathbf{E}) - \frac{i\kappa e\hbar^2}{2m}\left(\frac{\mathbf{a}\cdot\mathbf{x}}{c^2}\right)\beta\boldsymbol{\sigma}\cdot(\nabla\times\mathbf{E}),$$

$$\{\mathcal{O}_0,\mathcal{O}_1\} = (\mathbf{a}\cdot\mathbf{x})\pi^2 + 2c^2((\nabla\Phi_{\mathrm{G}})\cdot\boldsymbol{\pi}) - \frac{i\kappa e\hbar}{m}\beta(\mathbf{E}\cdot\boldsymbol{\pi}) - i\hbar(\mathbf{a}\cdot\boldsymbol{\pi}) \tag{D.28}$$
$$+ c^2\boldsymbol{\pi}\left(\frac{\mathbf{a}\cdot\mathbf{x}}{c^2}\right)\cdot\boldsymbol{\pi} - i\hbar c^2(\nabla^2\Phi_{\mathrm{G}}) - \frac{\kappa e\hbar^2}{2m}\beta(\nabla\cdot\mathbf{E}) + \hbar\boldsymbol{\sigma}\cdot(\mathbf{a}\times\boldsymbol{\pi}) -$$
$$-2e\hbar c\left(\frac{\mathbf{a}\cdot\mathbf{x}}{c^2}\right)\boldsymbol{\sigma}\cdot(\nabla\times\mathbf{A}) - \frac{i\kappa e\hbar^2}{2m}\beta\boldsymbol{\sigma}\cdot(\nabla\times\mathbf{E}),$$

$$[\mathcal{O}_0, [\mathcal{O}_1, \mathcal{E}_0]] = ie\hbar^2 \mathbf{a}\cdot\nabla\varphi - e\hbar^2(\mathbf{a}\cdot\mathbf{x})\nabla^2\varphi \\ -2e\hbar(\mathbf{a}\cdot\mathbf{x})\boldsymbol{\sigma}\cdot(\nabla\varphi\times\boldsymbol{\pi}), \tag{D.29}$$

$$[\mathcal{O}_0, [\mathcal{O}_0, \mathcal{E}_1]] = -4imc^2\hbar\beta(\mathbf{a}\cdot\boldsymbol{\pi}) - \hbar^2c^3\nabla^2(\nabla_0\Phi_G) \\ -i.e.\hbar^2c\,\boldsymbol{\sigma}\cdot(\boldsymbol{\omega}\times(\nabla\times\mathbf{A})) + 2mc^2\hbar\beta\boldsymbol{\sigma}\cdot(\mathbf{a}\times\boldsymbol{\pi}) \\ +4mc^2\beta(\mathbf{a}\cdot\mathbf{x})\pi^2 - 4emc\hbar\beta(\mathbf{a}\cdot\mathbf{x})\boldsymbol{\sigma}\cdot(\nabla\times\mathbf{A}) \\ -2\hbar c^3\boldsymbol{\sigma}\cdot(\nabla(\nabla_0\Phi_G)\times\boldsymbol{\pi}) + 2\hbar c^2\boldsymbol{\sigma}\cdot(\boldsymbol{\omega}\times\boldsymbol{\pi})\times\boldsymbol{\pi}, \tag{D.30}$$

$$[\mathcal{O}_1, [\mathcal{O}_0, \mathcal{E}_0]] = -\hbar^2ec^2\left(\frac{\mathbf{a}\cdot\mathbf{x}}{c^2}\right)\nabla^2\varphi - 2\hbar ec^2\left(\frac{\mathbf{a}\cdot\mathbf{x}}{c^2}\right)\boldsymbol{\sigma}\cdot(\nabla\varphi\times\boldsymbol{\pi}) \\ + 2\hbar ec^2\boldsymbol{\sigma}\cdot((\nabla\Phi_G)\times\nabla\varphi) - i\hbar^2e\,\boldsymbol{\sigma}\cdot(\mathbf{a}\times\nabla\varphi). \tag{D.31}$$

After neglecting the non-Hermitean terms in (D.27)–(D.31), it becomes evident that the low-energy approximation for the Dirac Hamiltonian is

$$H_{\mathrm{FW}} \approx mc^2\beta + \left[\frac{1}{2m}\pi^2 - \frac{e\hbar}{2mc}\boldsymbol{\sigma}\cdot\mathbf{B}\right]\beta - \frac{e\hbar}{4m^2c^2}\boldsymbol{\sigma}\cdot(\mathbf{E}\times\boldsymbol{\pi}) \\ - \frac{e\hbar^2}{8m^2c^2}[(\nabla\cdot\mathbf{E}) + i\boldsymbol{\sigma}\cdot(\nabla\times\mathbf{E})] + e\varphi \\ + \left[m(\mathbf{a}\cdot\mathbf{x}) - \frac{\kappa e\hbar}{2mc}\left(1+\frac{\mathbf{a}\cdot\mathbf{x}}{c^2}\right)(\boldsymbol{\sigma}\cdot\mathbf{B})\right]\beta \\ - \boldsymbol{\omega}\cdot(\mathbf{x}\times\boldsymbol{\pi}) - \boldsymbol{\omega}\cdot\left(\frac{\hbar}{2}\boldsymbol{\sigma}\right) + c(\nabla_0\Phi_G) \\ + \frac{1}{2m}\beta\boldsymbol{\pi}\left(\frac{\mathbf{a}\cdot\mathbf{x}}{c^2}\right)\cdot\boldsymbol{\pi} + \frac{\hbar}{4mc^2}\beta\boldsymbol{\sigma}\cdot(\mathbf{a}\times\boldsymbol{\pi}) - \frac{\hbar}{4m^2c^2}\boldsymbol{\sigma}\cdot(\boldsymbol{\omega}\times\boldsymbol{\pi})\times\boldsymbol{\pi} \\ - \frac{\kappa e\hbar^2}{4m^2c^2}\left(1+\frac{\mathbf{a}\cdot\mathbf{x}}{c^2}\right)[(\nabla\cdot\mathbf{E}) + i\boldsymbol{\sigma}\cdot(\nabla\times\mathbf{E})] \\ - \frac{\kappa e\hbar^2}{4m^2c^4}(\mathbf{a}\cdot\mathbf{E}) + \frac{e\hbar^2}{4m^2c^2}\left(\frac{\mathbf{a}\cdot\mathbf{x}}{c^2}\right)\nabla^2\varphi \\ + \frac{1}{m}\beta(\nabla\Phi_G)\cdot\boldsymbol{\pi} + \frac{e\hbar}{4m^2c^2}\boldsymbol{\sigma}\cdot\left(\nabla\varphi\times\left[(\nabla\Phi_G) + \left(\frac{\mathbf{a}\cdot\mathbf{x}}{c^2}\right)\boldsymbol{\pi}\right]\right) \\ + \frac{\hbar^2}{8m^2c}\nabla^2(\nabla_0\Phi_G) + \frac{\hbar}{4m^2c}\boldsymbol{\sigma}\cdot(\nabla(\nabla_0\Phi_G)\times\boldsymbol{\pi}). \tag{D.32}$$

The occurrence of non-Hermitean terms, here neglected, is a well-known phenomenon likely connected with the breakdown od the single-particle interpretation of the Dirac equation in the presence of time-dependent inertial and gravitational fields [8,9].

Equation (D.32) for the low-energy Hamiltonian can be now compared with the results obtained by other authors. Neglecting, for simplicity, the anomalous magnetic moment contributions, the Bonse–Wroblewski, Page–Werner, and Mashhoon terms can be immediately recognized by inspection. They correspond to the eighth, tenth, and eleventh terms, respectively. The fifteenth term contains electromagnetic and momentum corrections to the Mashhoon effect. The thirteenth term represents the redshift effect of the kinetic energy already mentioned, but here in the company of its electromagnetic corrections. These also appear in the new inertial spin–orbit term found by Hehl and Ni (the fourteenth term). The fourth and sixth terms represent spin–orbit coupling and are discussed, for instance, by Bjorken and Drell. The third term is also well known and represents the magnetic dipole interaction. The Darwin term is the fifth and the nineteenth represents an acceleration correction to it. All remaining terms are proportional to the derivatives of Φ_G (see Eq. (D.8)). Among these $c(\nabla_0\Phi_G) + \frac{1}{m}\beta(\nabla\Phi_G)\cdot\boldsymbol{\pi}$ appear to dominate. The integral-dependent part of (D.8) yields contributions that are small for small paths and low particle momenta. The largest contributions come from the first part of (D.8) which contains the terms $\frac{1}{2}mc^2\gamma_{00}(x)$ and $mc\gamma_{0i}(x)p^i$ already discussed by De Witt and Papini in connection with the behaviour of superconductors in weak inertial and gravitational fields.

In view of the above, Eq. (D.32) appears remarkably successful in dealing with all the inertial and gravitational effects discussed in the literature.

A derivation of a high-energy approximation can be found in the works of Cini and Touschek [10,11]. By proceeding in similar ways and again retaining only the leading order Hermitean terms, one arrives at the result

$$
\begin{aligned}
H_{CT} \approx\ & (1+\mathbf{a}\cdot\mathbf{x})\Bigg[\left(\sqrt{|\boldsymbol{\pi}|^2+m^2} - \frac{q^3}{\sqrt{1+q^2}}\frac{\hbar}{2m}\boldsymbol{\sigma}\cdot\mathbf{R}'\right)\frac{(\boldsymbol{\alpha}\cdot\boldsymbol{\pi})}{|\boldsymbol{\pi}|} \\
& +\frac{\kappa e\hbar}{2m}\beta\,(i\boldsymbol{\alpha}\cdot\mathbf{E}-\boldsymbol{\sigma}\cdot\mathbf{B})\Bigg] - \boldsymbol{\omega}\cdot(\mathbf{x}\times\boldsymbol{\pi}) - \frac{\hbar}{2}\boldsymbol{\sigma}\cdot\boldsymbol{\omega} + e\varphi + \boldsymbol{\alpha}\cdot\nabla\Phi_G + \nabla_0\Phi_G \\
& +\frac{q}{2|\boldsymbol{\pi}|}(1+\mathbf{a}\cdot\mathbf{x})\frac{\kappa e\hbar}{2m}\Bigg[\hbar\left(\nabla_k E^k\right) + 2\boldsymbol{\sigma}\cdot(\mathbf{E}\times\boldsymbol{\pi}) - \frac{2\hbar}{|\boldsymbol{\pi}|^2}\mathbf{R}'\cdot(\mathbf{E}\times\boldsymbol{\pi}) \\
& -2\left(1-\frac{\hbar}{|\boldsymbol{\pi}|^2}\boldsymbol{\sigma}\cdot\mathbf{R}'\right)\mathbf{B}\cdot\boldsymbol{\pi}\Bigg] \\
& +\frac{q}{2|\boldsymbol{\pi}|}\left[-\frac{\hbar}{|\boldsymbol{\pi}|}\sqrt{|\boldsymbol{\pi}|^2+m^2}\left(1-\frac{\hbar}{|\boldsymbol{\pi}|^2}\boldsymbol{\sigma}\cdot\mathbf{R}'\right)\sigma^k\,\epsilon^i{}_{jk}\,\nabla_i\,(\mathbf{a}\cdot\mathbf{x})\,\pi^j\right. \\
& +\frac{\hbar^2}{|\boldsymbol{\pi}|^2}\boldsymbol{\alpha}\cdot\left[\mathbf{R}'\times\nabla\left((\boldsymbol{\omega}\times\mathbf{x})\cdot\boldsymbol{\pi}\right)\right] - \frac{\hbar^2}{2}\alpha^k\,\epsilon^i{}_{jk}\left(\nabla_i\omega^j\right) \\
& +\frac{\hbar^2}{|\boldsymbol{\pi}|^2}\left[(\mathbf{R}'\cdot\boldsymbol{\omega})\,\boldsymbol{\alpha}\cdot\boldsymbol{\pi} - \alpha^j\,R'^k\,\omega_j\,\pi_k\right] \\
& \left. +\,2\left(1-\frac{\hbar}{|\boldsymbol{\pi}|^2}\boldsymbol{\sigma}\cdot\mathbf{R}'\right)\left(\nabla\Phi_G\cdot\boldsymbol{\pi} - \frac{\hbar^2}{2}\left(\nabla_k a^k\right)\right)\right]\beta,
\end{aligned}
\tag{D.33}
$$

where $R'^k = R^k - e\,\epsilon^{ki}{}_j\left(\nabla_i A^j\right)$.

Fermion Helicity Flip in Weak Gravitational Fields

E

The helicity flip probability is calculated in this appendix, at the tree level, for a Dirac particle that interacts via a graviton exchange with a massive scalar particle in Minkowski space–time. As usual, the space–time metric is written as $g_{\mu\nu} = \eta_{\mu\nu} + \kappa h_{\mu\nu}$, where the perturbation $h_{\mu\nu}$ represents the graviton field (which is supposed to be quantized in Minkowski space–time).

Following [12–14], one finds that

$$V^{\mu\nu}_{\phi\phi h}(k_1, k_2) = i\frac{\kappa}{2}\left[k_1^\mu k_2^\nu + k_1^\nu k_2^\mu - \eta^{\mu\nu}(k_1 \cdot k_2 + m^2)\right], \tag{E.1}$$

and

$$V^{\mu\nu}_{\bar{\psi}\psi h}(k_1, k_2) = i\frac{\kappa}{4}\left\{\frac{1}{2}\left[\gamma^\mu\left(k_1^\nu + k_2^\nu\right) + \gamma^\nu\left(k_1^\mu + k_2^\mu\right)\right] + \eta^{\mu\nu}\left[(\not{k}_1 + \not{k}_2) - 2m\right]\right\}. \tag{E.2}$$

Consider now the scattering of a fermion F interacting with a scalar S via a graviton exchange, $F(p, L) + S(q) \to F(k, R) + S(l)$ (see Fig. (E.1)), where L (R) stands for the left-(right-)handed helicity. The helicity flip rate is given by [12]

$$N_{L\to R} = \left|\bar{u}(k, \mathcal{S}_R)\, V^{\mu\nu}_{\bar{\psi}\psi h}(p, k)\, u(p, \mathcal{S}_L)\, D_{\mu\nu,\alpha\beta}(p-k)\, V^{\alpha\beta}_{\phi\phi h}(q, -l)\right|^2, \tag{E.3}$$

where $D_{\mu\nu,\alpha\beta}$ is the graviton propagator

$$D_{\mu\nu,\alpha\beta}(p-k) = \frac{i}{2(p-k)^2}\left(\eta_{\mu\alpha}\eta_{\nu\beta} + \eta_{\mu\beta}\eta_{\nu\alpha} - \eta_{\mu\nu}\eta_{\alpha\beta}\right), \tag{E.4}$$

$\mathcal{S}^\mu_{L:R}$ are the polarization four vectors

$$\mathcal{S}^\mu_L = -\frac{p^\mu}{m\beta_i} + \frac{\sqrt{1-\beta_i^2}}{\beta_i}\,\eta^{\mu 0},$$

G. Lambiase and G. Papini, *The Interaction of Spin with Gravity in Particle Physics*, Lecture Notes in Physics 993, https://doi.org/10.1007/978-3-030-84771-5_E

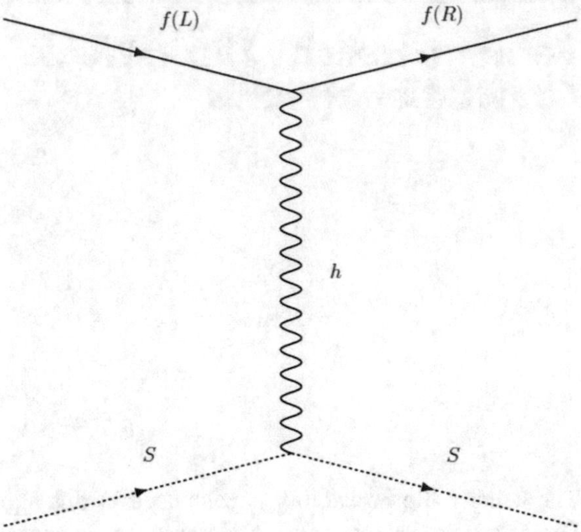

Fig. E.1 Scattering $f(p, L) + S(q) \to f(k.R) + S(l)$, where L, R indicate the left- and right-handed helicity

$$\mathcal{S}_R^\mu = \frac{k^\mu}{m\beta_f} - \frac{\sqrt{1-\beta_f^2}}{\beta_f}\eta^{\mu 0} \tag{E.5}$$

with $\beta_{i(f)}$ the initial and final fermion velocities. The squared invariant amplitude for an initial left-handed fermion $|L\rangle$ that can flip to a right-handed one $|R\rangle$ (due to a graviton exchange with a scalar field) is

$$N_{L\to R} = \frac{\kappa^4}{512t^2}\Big\{\left[1-(S_L\cdot S_R)\right]N_1 - 4\left[(A\cdot S_R)(k\cdot S_L)+(A\cdot S_L)(p\cdot S_R)\right]N_2 \\ + 2(p\cdot S_R)(k\cdot S_L)N_3 + 8(A\cdot S_L)(A\cdot S_R)t\Big\}, \tag{E.6}$$

where

$$N_1 = (s-u)^2\left[(s-u)^2 - t(t-4M^2)\right] - 16m^2M^2\left[M^2(t-4m^2)+(s-u)^2\right],$$
$$N_2 = (s-u)^2 + 2t(t+3M^2) - 8m^2M^2,$$
$$N_3 = 4(t+3M^2)^2(t-4m^2) + (s-u)^2(5t+8M^2) + 16m^2(t+2M^2)^2,$$

while $s = (p+q)^2$, $t = (p-k)^2$, and $u = (p-l)^2$ are the Mandelstam variables, and $A^\mu \equiv (t+3M^2)(k^\mu + p^\mu) - \frac{1}{2}(s-u)(q^\mu + l^\mu)$. A similar expression holds for the amplitude describing the process in which a initially left-handed fermion preserves the same helicity after the interaction.

One may then introduce the polarization of the scattered fermion, defined as

$$P = 1 - \frac{2N_{L\to R}}{N_{L\to L} + N_{L\to R}}, \tag{E.7}$$

where the quantity

$$N_{L\to L}+N_{L\to R}=\frac{\kappa^4}{256t^2}\left\{(s-u)^2\left[(s-u)^2-t(t-4M^2)\right]\right.$$
$$\left.-16\,m^2M^2\left[M^2(t-4\,m^2)+(s-u)^2\right]\right\} \tag{E.8}$$

depends only on the Mandelstam variables and the relevant masses. Notice that only massive fermions can have their helicity flipped. The value $P=1$ means that there is no depolarization (flip) of the initial fermions, while the value $P=-1$ means that the left-handed initial fermions are flipped. In the limit of very large scalar masses, (*i.e.* $M \gg E > m$), that is, in the small squared momentum transfer approximation, one finds that the helicity flip probability simplifies to

$$P \simeq 1-\left(\frac{m^2}{E^2}\right). \tag{E.9}$$

As (E.9) shows, P is proportional to the square of the fermion mass. The $\frac{m}{E}$ ratio is very small for relativistic fermions. This implies $P \approx 1$, that is, the helicity of the fermion remains unchanged.

A similar calculation can be carried out for the torsion field [15].

References

1. Kh, E., Akhmedov and A. Yu. Smirnov, : Phys. At. Nucl. **72**, 1363 (2009)
2. Kh, E., Akhmedov and A. Yu. Smirnov, : Found. Phys. **41**, 1279 (2011)
3. Cardall, C.Y., Fuller, G.M.: Phys. Rev. D **55**, 7960 (1997)
4. Fornengo, N., Giunti, C., Kim, C.W., Song, J.: Phys. Rev. D **56**, 1895 (1997)
5. Swami, H., Lochan, K., Patel, K.M.: Phys. Rev. D **102**, 024043 (2020)
6. Foldy, L.L., Wouthuysen, S.A.: Phys. Rev. **78**, 30 (1950)
7. Bjorken, J.D., Drell, D.S.: Relativistic Quantum Mechanics. McGraw-Hill, San Francisco (1964)
8. Parker, J.L.: Phys. Rev. D **22**, 1922 (1980)
9. Huang, J.C.: Ann. Physik **3**, 53 (1994)
10. Cini, M., Touschek, B.: Nuovo Cim. **7**, 422 (1958)
11. Singh, D., Mobed, N., Papini, G.: J. Phys. A **37**, 8329 (2004)
12. Aldrovandi, R., Matsas, G.E.A., Novaes, S.F., Spehler, D.: Phys. Rev. D **50**, 2645 (1994)
13. B. S. DeWitt, Phys. Rev. **162**, 1239 (1967). D. J. Gross and R. Jackiw, Phys. Rev. **166**, 1287 (1968)
14. V. I. Ogievetskii, and I. V. Polubarinov, Soviet Phys. JETP **21**, 1093 (1965). S. F. Novaes, and D. Spehler, Phys. Rev. D**44**, 3990 (1991)
15. SenGupta, S.: A, Sinha. Phys. Lett. B **514**, 109 (2001)

Index

A
Axions, 143

B
Berry, 1–3, 6, 9, 143
Bose–Einstein condensates, 9

C
Chirality, 29
Compound spin, 29, 31, 37, 39, 41
Covariant wave equations, 1

D
Dirac, 1, 2, 6, 12, 14, 17, 18, 20
Dirac matrices, 14
Discrete symmetries, 33, 35
Double copy, 165

E
Equivalence principle, 1, 12

F
Fierz–Pauli equation, 20
Flux, 52, 55

G
Gauge transformations, 12, 22
Geometrical optics, 75, 77, 90
Gravitational lensing, neutrino, 109
Gravitational memory effect, 151

H
Hein-Fritz London, 144
Helicity flip, 77
Helicity oscillations, 71
Hilbert space, 2, 6
Holography, 164

I
Interferometers, 51, 53, 55, 59, 65
Invariance, 30, 31

K
Kawai–Lewellen–Tye relations, 164
Klein–Gordon, 1–4, 6, 8, 18

L
Lanczos–DeDonder, 9, 11, 21
Lense–Thirring effect, 51, 59, 60, 64
Lensing, 60, 61, 63, 64
Loop Quantum Gravity, 165

M
Mashhoon, 29, 30, 33
Mashhoon effect, 30–32
Meissner effect, 144
Muon $g-2$ Collaboration, 36

N
Neutrino, flavour oscillations, 101, 102
Neutrino, magnetic momentum, 101, 102, 107
Neutrino, spin-flip, 101

G. Lambiase and G. Papini, *The Interaction of Spin with Gravity in Particle Physics*, Lecture Notes in Physics 993, https://doi.org/10.1007/978-3-030-84771-5

P
P invariance, 29, 33

Q
Quantum phases, 2, 6, 20
Quasi-normal modes, 151, 159

R
Radiative processes, 114, 120
Ricci, 14, 17
Ricci scalar, 19
Ricci tensor, 21
Riemann tensor, 22

S
Spin connection, 15
Spin currents, 113, 123–125
Spin-gravity, 2, 22
Spin memory effect, 151
Spin-rotation coupling, 37
Spin-2, 1, 20, 23
String theory, 165
Superconducting interferometers, 51
Superconductors, 8
Superfluid interferometers, 55
Superfluids, 8, 9

T
Tetrad, 13, 16, 17, 19
T invariance, 29, 33
Torsion, 141–143

V
Vierbein, 12–16, 19
Vortices, 113, 125–130

W
Wave function, 2, 22
Wave optics, 60, 64
Weyl, 19

Printed in the United States
by Baker & Taylor Publisher Services